BELL LABS MEMOIRS:
VOICES OF INNOVATION

Cover photograph: The glass telephone blown by David Dorsi.
(Photograph reprinted by permission of David Dorsi.)

A product of the IEEE History Center

ISBN-10: 1463677979
ISBN-13: 9781463677978

BELL LABS MEMOIRS: VOICES OF INNOVATION

Edited by

A. Michael Noll &
Michael N. Geselowitz

TABLE OF CONTENTS

FOREWORD

Jeong H. Kim
Executive Vice President, Alcatel-Lucent, and
President, Bell Labs

There are a only a handful of scientific research institutions on this planet that simultaneously infuse their
members with a dedication to discovery, confer upon
them a legacy of technological breakthroughs, and
nurture among them a culture of problem-solving
and the deepest mutual respect. One such institution is Bell Laboratories. Dr. William O. Baker
led the research division of Bell Labs and in many
respects was responsible for the qualities that made
it great. The institution and the man thrived on
each other, grew together, and maintained bonds
that transcended life itself: Dr. Baker died in 2005
at the age of 90, but his impact on our work is very
much alive. This book – though not directly about

Dr. Baker — attempts to capture the essence of that research institution through the lives of a collection of people who worked there during Baker's tenure at its helm.

The Bell Labs research organization that Dr. Baker led — as Vice President of Research from 1955 to 1973 and as President from 1973 to 1979, retiring as Chairman a year later — achieved an astonishing record of invention. Even the most abbreviated compilation would include the discovery of radio astronomy, the transistor, cellular communications, information theory, the solar cell, the laser, speech synthesis, UNIX, the charged coupler device used in so many of today's cameras, and the C programming language. Under his leadership, Bell Labs established a reputation that remains powerful to this day, as a transformative agent that has driven the evolution of science, information networking, and the communication bonds that serve humankind.

In the pages that follow, people who worked at Bell Labs during Baker's time provide an insight into the gestalt of that institution and how it enriched the world in which they lived. But I would like to preface that story by introducing the reader to the Bell Labs of today. I make these comments as a foreword rather than an epilogue to provide context for readers as they contemplate the evolution of Bell Laboratories after the breakup of the old Bell System in 1984.

Lately, some industry observers have incorrectly portrayed the Labs as a relic; they think the industrial research laboratory has gone the way of the manual typewriter and the slide rule. On the contrary, the Bell Labs that emerged after the Bell breakup has grown more relevant, not less, and even today offers promise to an industry facing radical change and technological disruption. Other information and networking enterprises have also recognized that industrial research in a turbulent and rapidly changing industry is a necessity that cannot be fully outsourced to academics or other institutions. Companies such as Microsoft, Google, IBM, HP, and Hitachi – to name just a few – clearly see the relevance of industrial research to their future growth and have either sustained or created such organizations. William Baker understood this dynamic and established the priorities that enable Bell Labs to continue to play a vital role in the industry. His accomplishments made it far easier for subsequent leaders – myself included – to sustain the unique position that Bell Labs holds today.

Above all, Baker recognized that it is the quality of our individual researchers that underlies the strength of our collective institution:

> It is that the person makes the place. Brains, character, integrity, energy, loyalty – the things that make fine people – make fine engineering and fine science for the Bell

System. Those who support ... all these efforts toward excellence share equally in this principle that the quality of our people is the essence of our progress. (Remarks of William Baker, Bell Labs Cabinet Meeting, New York City, January 11, 1974)

Today's Bell Labs holds on to that precept, seeking out the pre-eminent research scientists in a diverse set of research disciplines and providing them with an environment where they can work together to pursue discoveries of a disruptive nature, even as they absorb, interact with, and then incorporate the concrete and near-term needs of customers. These attributes of current research at Bell Labs – fundamental research disciplined by the strictures of real-world application – have been at the heart of Bell Labs from its inception, and they were actively cultivated by Bill Baker.

The distinction between the academic and industrial research environments – the urgency to be responsive to customer needs in an accelerated timeframe – remains of paramount importance to present-day Bell Labs, as it was to Dr. Baker:

Excellence in research and development, particularly in industry, has rather different parameters than the usual qualities of excellence in scholarship, in learning, in art, in the humanities, and even in business and

4

public affairs. Namely, it has to be mastered and applied in an environ in which speed of discovery, recognition of broad new outlines of knowledge, rapid and early innovation, and record of findings are dominant features. (Remarks of William Baker, Centennial Lecture at Arizona State University, Tempe, Arizona, January 1988)

As one might expect, there are some differences between the research at Bell Labs led by William Baker and the Bell Labs of today. The changing character of the parent company, from regulated monopoly to competitive enterprise, saw the elimination of several departments that fell outside our core competencies in communications technologies (departments such as our world-renowned economics group). That said, Bell Labs retains its highly regarded mathematical and algorithmic sciences team). There has also been a shift in the allocation of resources from the physical to nonphysical sciences, although we retain a strong capability in technologies such as nanotechnology, optics, and radio sciences. And Bell Labs is a culturally and geographically more diverse organization than it once was, reflecting the fact that pre-eminent researchers are now more evenly distributed around the globe.

Perhaps the most significant difference is our focus on coupling innovation with the needs of the marketplace. In today's world, where the seconds on the clock seem to tick off a little faster than in the past, we have gone to great lengths to balance the creation of what we call disruptive research assets and inventions with the need to rapidly transition them into the market.

This balancing act is a challenge, because the times in which we live and the markets in which we compete demand far more than incremental improvements. This is an era for grand challenges, a generation that demands of us inventions that might have been considered the realm of science fiction just a few years ago. But at the same time, we must apply our energies and passions to innovations that will have immediate impact, perhaps within a year or two. Thus, we began a ventures initiative in which unique and promising research, not yet matured for commercial development by business groups can be funneled into an incubator where researchers continue to invent while working directly with potential customers. These Bell Labs researchers are subsequently forged into small ventures teams – supplemented with business experts and technicians – to accelerate the commercialization of their research endeavors.

But let me conclude by celebrating one of the constants that unites William Baker's era with our own. Whether the Bell Labs researcher is contemplating the theoretical basis for quantum computing, exploring ways to convert radio interference into more bandwidth, or working with security managers on a new platform for safeguarding laptops, that researcher makes a commitment to technical excellence, creativity, and innovation—to the principles that defined Bell Labs science under Dr. Baker and drive our aspirations to this day. Dr. Baker and all of the presidents of Bell Laboratories – those who preceded him and those who followed him – recognized the importance of sustaining these principles against the onslaught of day-to-day administrative challenges, the temptations to take short cuts, and the perils of short-termism. Those of us privileged to preside over this great institution draw continuing strength from our predecessors and the principles they upheld. We honor their stewardship by seeking to emulate it in the years ahead.

PREFACE

In October 2005, Dr. William Oliver Baker died at a nursing home in New Jersey. Relatively unknown on a national scale, Baker headed the Research Division at Bell Labs from the latter half of the 1950s to the early 1970s – what is acknowledged to have been one of the golden ages of research at Bell Labs. But far beyond his impressive leadership at Bell Labs, Baker was extremely active in consultation and national politics. He headed numerous boards and committees in the non-profit sector and was a personal advisor to many influential people in the United States, including every president from Eisenhower to Reagan, for each of whom he served as an unofficial advisor on matters of communications and national security. His death signaled both the end of an era in industrial research and of the concept of citizen diplomacy in the arena of national security.

Before his death, Dr. Baker enlisted me to organize his papers for donation to Princeton University, from which he received his doctorate in chemistry in 1939. In the process of working closely with him on this archival project, I was afforded the opportunity to interview him on a host of subjects, many of which are represented in this book. (Additional information about Dr. Baker, including transcripts of some of these interviews, is posted at williamobaker.org.)

The real challenge, in preparing this book, was how to memorialize this man and his impressive achievements, since Dr. Baker never wanted attention and glory for himself and always deferred to his beloved Bell Labs and the people there. Although many people encouraged him to write an autobiography or memoir, he humbly refused, stating that no one would be ever interested in reading it. During interviews, he always politely resisted speaking about his childhood or parents, preferring to preserve a sense of personal privacy. What all this meant is that a book or event focused on Baker possessed all the inherent difficulties of paying just homage to a man who sought no such distinctions.

Nevertheless, the solution was simple: follow Baker's own lead. As such, this book places the focus precisely where Baker would have: on the lives of the people who made Bell Labs during the time when he headed the research organization at Murray Hill, NJ. Some contributed a memoir or an autobiography;

others offered their reflections in interview. In consonance with Baker's respect for people at all levels and tasks, those asked to present their lives in this book also cover a broad spectrum of work at Bell Labs.

There were hundreds of people who worked at Bell Labs in Baker's research area during the 1950s and 1960s. Some of these people I knew personally from my own employment there during the 1960s in the acoustics department. One of the challenges for this book was to choose a small but representative sample of the people who were involved with Baker during his years in leading the research organization. Personal judgment played an inevitable role. My objective was to assure diversity in the backgrounds of the people rather than attempt to cover all the various disciplines and departments that comprised Baker's research area. This approach seems fitting, as Baker was particularly proud of the diversity of the people in research, including many who had been recruited from overseas.

Baker cared much about history and getting the facts correct. He was deeply involved with the IEEE History Center. Accordingly, this book has made extensive use of the Center's archives and has been co-edited by its director, Dr. Michael Geselowitz. Ellen T. Connett of Bell Labs was responsible for communication with all the participants in this project and for keeping everything well organized and reasonably on schedule.

The interviews of Duke Dorsi and Italo Quinto used in this book were conducted by Richard Q. Hofacker, who used to work in media relations at Bell Labs. His efforts in making this project a reality are gratefully acknowledged, as are all those who contributed their stories to this book. My co-editor, Michael Geselowitz, and I would also like to acknowledge the work of IEEE History Center staff members Robert Colburn, Sheldon Hochheiser, Matthew Friedman, and Steven McGrail in preparing this book for publication, and Ed Eckert, Archivist at Alcatel-Lucent for his help in supplying many of the photographs which illustrate it.

A. Michael Noll
Stirling, NJ

Michael N. Geselowitz
New Brunswick, NJ

INTRODUCTION

A. Michael Noll

Overview

Bell Telephone Laboratories in the 1950s and 1960s was lauded as the world's greatest industrial research organization. Indeed, today's digital era had much of its genesis in research performed at Bell Labs then, and even earlier, from digital computers and information theory to digital music, art, and animation.

    William O. Baker was Vice President, Research, at Bell Labs from 1955 to 1973. Baker's philosophy in guiding the research area was based upon his predecessor and mentor, Ralph Bown. In 1954, Bown presented a paper about how to manage a research organization at the Sixth Annual Conference on the Administration of Research.[1] Bown's paper is specific in its advice and clearly had significant impact on Baker and how he would manage research during his tenure at Bell's helm. Baker and Bown were good personal friends and sailed together on vacations.

The challenge was how to capture the spirit of Bell Labs with an emphasis on the people who contributed to it, particularly when thousands were involved. Baker had a deep curiosity about the backgrounds of the people who worked at Bell Labs – their family lives, education, and accomplishments – and he always remembered their names and histories. Baker also had a strong affection for the people who made Bell Labs and would always place the emphasis on them and not on him.

This suggested an approach to this book – namely, to present a snapshot of the history of Bell Labs through the personal memoirs, autobiographies, interviews, and biographies of a small sample of the people who worked at Bell Labs in the 1960s during Dr. Baker's reign in the research organization. Although the focus is on the people who worked in research at the Murray Hill, NJ facility, a selection of

---

1 Ralph Bown, "Vitality of a Research Institution and How to Maintain It," Proceedings of the Sixth Annual Conference on Administration of Research, *Bell Telephone System Monograph* 2207, 1953.

support staff is also included both to present a broader perspective and also because Baker had the deepest respect for people at all levels and tasks at Bell Labs.

In presenting these personal memories of Bell Labs, the book not only does honor to William O. Baker; it also documents the human element that is so essential to creating a great industrial research facility.

PARTICIPANTS

In considering the participant base for this book, diversity was a prime consideration. In an interview in 2002, Baker described how a conscious effort was begun in the late 1950s and 1960s to recruit and appoint "a very appreciable number of very promising people from communities overseas that had not been connected with the Bell Labs before." He remarked that "Bell Laboratories was notorious for not having done that over the years before this." So in this book we have people from overseas: Alan Chynoweth (Britain), Manfred Schroeder (Germany), and Mohan Sondhi (India). The diversity extends to including support personnel, such as Duke Dorsi (glassblower), William Keefauver (patent attorney), and Italo Quinto (limousine driver). Baker did not care about a person's job status and treated everyone the same.

John R. Pierce is the first person treated in this book. Pierce was the father of communications satellites and was responsible for both *Echo* and *Telstar*.

He headed a research division under Baker and was a powerhouse of ideas and well-posed challenges to his researchers. Born in Iowa of relatively poor parents, he spent his early years there before, in his teens, the family moved to southern California where he ultimately obtained bachelors and doctoral degrees in engineering from Cal Tech. Pierce went immediately to Bell Labs and initially worked on klystron vacuum tubes; his story concludes with reference to Dr. Ralph Bown, Baker's immediate predecessor as Vice President for Research.. Throughout his account, Pierce discusses some of the signal factors that contributed to the success of Bell Labs, most notable of which was the unique community atmosphere fostered there.

Manfred Schroeder came to the Labs from Germany as a physicist and made significant contributions to speech encoding with his innovations of various Vocoder techniques and methods for speech pitch detection. He is also well respected for his research into room and concert hall acoustics, including early work on the New York Philharmonic. Other workers in Schroeder's department pioneered the use of thin, permanently charged foils as condenser microphones, known as electret microphones, which today are used in nearly every cell phone and many other consumer electronic devices.

Virginia native Walter Brown began his education in physics in the Navy during World War II and came to the Labs in 1950 fresh off his doctoral work at Harvard. Specializing in nuclear physics, he performed research in the properties of germanium, ion implantation, and the development of silicon semiconductors. In the early '60s, Brown became intimately involved in designing the pioneering *Telstar* communication satellite.

Carol Maclennan worked for nearly four decades at Bell Labs as a computer programmer, in the process becoming one of the first women to ascend from the level of Technical Assistant to full Member of Technical Staff. Her labors at Bell Labs included work on the *Ulysses* spacecraft and brought her to such remote research sites as Greenland and Antarctica.

Alan Chynoweth grew up on the south coast of England and World War II stimulated his interest in science and technology. He studied physics at King's College London. He went to Canada on a post-doctoral fellowship in 1950 and from there to Bell Labs in 1953 to work in physics research and later to spearhead the Labs' Metallurgical Research Laboratory. Chynoweth specialized as well in Bell Labs' extensive foreign recruitment program during the 1960s.

Duke Dorsi, a Stirling, New Jersey native, spent forty years at Bell Labs as an expert glassblower. His relationship to Dr. Baker extended beyond the Labs to hunting woodcock in the wilds of New Jersey. Dorsi relates numerous anecdotes about the climate of Bell Labs, and the interrelation of technical and support staff, during its "golden age."

Edward E. Zajac, whose early work for Bell Labs included laying and recovery of submarine telegraph cables, specialized in computer animation and economics. A Clevelander by birth, Zajac eventually became head of the Bell Labs economics research group, which, designed initially as a legal defense team, soon reached a level of expertise comparable to the finest university economics departments.

Edwin Chandross grew up in Brooklyn, New York and obtained his doctorate from Harvard. He came to Bell Labs in 1959 to work in the field of chemistry and his researches ranged from optical memory to the use of organic materials in electronics. Chandross was instrumental in exploring the chemistry of chemiluminescence, the fundamental principle of the ubiquitous light stick.

Italo Quinto, born in Lenola, Italy, found his way to Bell Labs as a bus and truck driver in the 1960s. His story travels from service as a gunner during World War II to that of salaried executive chauffeur, eventually serving as Dr. Baker's personal driver.

Mohan Sondhi came to the Labs in 1962 and was a co-inventor of the adaptive echo canceller used today in all long-distance voice transmission. Sondhi grew up in Delhi and after studying physics and engineering in Delhi and Bangalore came to the University of Wisconsin for his graduate studies. His first published paper appeared while he was still en route.

Patent attorneys are a unique breed, combining

competence in both engineering and law. William Keefauver grew up in Gettysburg, Pennsylvania and attended Penn State, only to find his undergraduate studies in electrical engineering interrupted by the Second World War. After serving in the Army Air Corps, he joined the patent division of Bell Labs, was sent to NYU's Law School to obtain a law degree, and thereafter ascended to the position of general counsel and eventually General Patent Attorney in charge of Bell's entire patent operation in the 1970s-80s. Keefauver's work for Bell was supplemented, during this period, by extensive diplomatic work in the arena of international patent negotiation.

The last person profiled in this book is William O. Baker, for whom this book is intended as a memorial tribute. Baker was Vice President, Research, at Bell Labs during the late 1950s and the 1960s, when all the people profiled in this book were at Bell Labs, either working in his research area or closely with him. Baker clearly had been groomed by management to take over the research helm at Bell Labs. While his predecessor, Ralph Bown, had already set the tone and laid out the principles for the management of research at Bell Labs, Baker went far beyond Bown's blueprint and set such policies as the recruiting of scientists from overseas at a time when its parent organization, AT&T, questioned such hiring. Baker also defended research into computer music, art, and animation from strong criticism by AT&T. Initiating programs for the advanced education of employees and the promotion of women and minorities, Baker was a true visionary who had considerable impact on both the culture of Bell Labs as well as its illustrious research efforts.

As a fitting conclusion to the overall theme of this book, its Epilogue, by Michael Geselowitz, discusses the broader impact of Bell Labs and what he calls "the Baker period." He also discusses the patterns that stretch across the individual narratives in the book, such as the loyalty of the individuals to Bell Labs, their ethnic and educational diversity, a bottom-up management structure, and lastly the personal hand of Bill Baker on the research organization.

JOHN R. PIERCE

John R. Pierce.
(Reprinted with permission of Alcatel-Lucent USA Inc.)

Note: This autobiography was edited from a much longer work *My Career as an Engineer: An Autobiographical Sketch by John R. Pierce* published by the University of Tokyo in honor of Pierce receiving the Japan Prize. The original work, dated September 22, 1985, was copyrighted by Pierce in 1988 and is edited and reproduced here with the permission of his widow, Brenda Pierce. The editing by A. Michael Noll consisted of deleting detailed material about people and the technical principles of his research – not a single word of Pierce's was changed, nor anything added. Pierce passed away on April 2, 2002, a victim of Parkinson's disease.

Another World

I tend to think little of the past or future; chiefly I live in the present. I have never kept a diary. I don't keep personal letters. The only written record I have of my most personal feelings is a few dozen poems that I have written over many years.

Technology changes the world we live in and the sorts of lives we lead. We may argue whether the changes are good or bad, but that there is change and that the change affects us is incontrovertible.

More often, technological change is so gradual that it escapes notice. The cars we drive and the highways we drive them on have changed much since the day of the first Model T Ford. Yet, we are more concerned with choosing a new car than with the revolution that automobiles and highways have worked in our lives.

Our world is profoundly different from that of our fathers, or, from that of our childhood. It is only by thinking back that we can see that this is so.

In 1910 I was born into a world quite different from ours. That world, its people, its way of life have been destroyed, not by time alone — for time kills only individuals — but by the science and technology that individuals, living and dead, have created.

The difference between the world of my childhood and our world is far greater than the absence of TV, or even of radio. In the home of my father's parents in Williamsburgh, Iowa, there was neither gas nor electricity. When we visited there, my mother heated her curling iron by hanging it down the glass chimney of a kerosene lamp. My grandmother cooked meals on the black cast-iron wood-burning range in the kitchen, or on a kerosene stove when there was no fire in the range. She pumped water into the sink from a cistern. I still remember the squatty pump with the iron handle and broad spout. The toilet was an outhouse.

Though I was born in Des Moines, my parents left that city when I was an infant. My earliest recollections are of Cedar Falls, where we stayed with my grandparents, and of St. Paul, Minnesota. There we lived at first in a huge block of flats, side by side and several stories high.

Later we moved to a duplex on Grand Avenue, farther from the center of the city. A few years after that my parents bought a bungalow at 2100 Grand Avenue.

Because I was born in 1910, I know the part that the first World War played in American life. At the movies (silent, of course) we saw news reel shots of marching men. We saved tinfoil (really tin, then, and not aluminum) and turned it in for some obscure government use. We saved peach pits and turned them in to make charcoal for gas masks. My mother made me a soldier suit. We sang war songs and hated Kaiser Bill. German candy stores were supposed to put ground glass in candy, and they closed. The Heidelberg restaurant in St. Paul changed its name. Hamburgers became Liberty steaks. The whole family got the Spanish influenza, and we were very ill.

Finally, the war was over. About 1919 my parents bought their first automobile. We began to go to places that had been inaccessible before. We traveled over a maze of primitive roads, gravel and macadam, unmarked, unnamed, unnumbered.

The automobile and the telephone worked great changes in American life. People built houses, and later shops, away from the streetcar lines. People called friends up before dropping in — though dropping in has persisted as an American custom.

However nostalgic I may be about the world of my childhood, it is gone, and so are the sorts of people who lived in it. Science and technology destroyed that world and replaced it with another. Now, people as well as things are different.

How did those changes come about? Through what men did; I among them. Technology has no life or end of its own. Men create technology in order to use it. I was drawn into the process of change through other men: sometimes I was influenced by men whom I knew, sometimes by what men wrote. In a large part, whatever I have done, whatever I am, reflects the personalities and the works of others even more than it reflects that strange, strange world I recall as I write these lines.

Growing Up

From my father I inherited or acquired a short chuckle or snort of amusement, though I am less good-humored and less often appreciative than he. He was a good, tolerant, patient, cheerful, hardworking man. It is a tribute to his integrity that my maternal grandfather, who had four married daughters, named my father as executor of his estate.

I think that in many ways my father and I were closer than it seemed. When I was staying with him during his late seventies, I told him that sometimes I lay awake at night puzzling over technical problems. He said that before he made some improvement in his cottage at Newport Beach he had spend wakeful hours considering how he might go about it. He asked if this was like me and my technical puzzling. I said that it was very like.

When I was a child, I would have been closer to my father had I seen more of him. He was away for weeks at a time, traveling with trunks of samples, selling ladies hats to the proprietors of millinery stores in small towns in Minnesota, South Dakota and Iowa.

I was an only child and my mother's boy. She was livelier than my father, a short, slender woman among three tall sisters, but nearly as tall as my father, and taller in high heels. Two of her sisters had gone to Normal School at Cedar Falls, now grown to a branch of the University of Iowa, and had taught school before marrying. Another sold insurance. My mother went to far Des Moines, became head trimmer at Ike Stern's millinery house, and there married my father in 1908, at the age of thirty-five.

My mother had a mechanical bent. She needed it to cope with the house while my father was away. I remember a picnic at Indian Mounds Park in St. Paul, for the employees and families of Strong and Warner's wholesale millinery house. There were games-sack races, and a nail-driving contest for the ladies. My mother won, although she broke the toy hammer she had been given and had to finish driving the nail with the heel of her shoe.

At home I learned to use the old treadle sewing machine. I remember vividly the bright spot of blood where the needle punctured my fingernail. But I also played with tinker toys, and later with my American Model Builder, an extinct relative of the Erector and the Meccano. I loved to make machines that operated in some way, and especially one that drew complex designs on a disc of paper.

I was fascinated by toy steam engines — whose principle I divined — by magic lanterns, which I took for granted, and by toy electric motors, which I misunderstood. I did acquire a Meccano electric motor with changeable gears, and I used it for many years to drive various contraptions. I have succeeded in using things even when I didn't understand them. I have found this to be the common way — the only way — of getting through life.

I wanted to hear about technical things even if I didn't understand them. As I remember it, my mother got technical books from the library even before I could read (I didn't' learn to read until I was seven), and read them to me. I was fascinated by such mysterious terms as electromotive force.

Perhaps I was slow to learn to read because I was young for my age in every way, and have been so always. I went to school late. I was a frail child, and never found it easy to face the world or others. At the usual age of six I was sent for a time to a school in a huge brick building. The large classes and the orderly procedures upset me terribly. I wonder what would have happened had I been sent by bus to some still larger and "better" school.

Instead, I was kept home for another year. I then went to first grade at a portable schoolhouse.

It had been set up on open land our home, to serve an outlying neighborhood during the war. To me, the pupils in the regional school seemed bored and apathetic compared with lively boys and girls of all ages in the country school.

In my poor portable school I had a wonderfully sympathetic teacher, Miss McVey. Because of her I was soon taught to read. Then, during recess periods, I was asked, along with other bright pupils, to help those who found reading difficult. In time I skipped a grade, and almost caught up, though I was always a little older than most fellow students.

My mother encouraged me in all sorts of technical play, before I went to school and after. I installed a single headlight on the front of my pedal automobile, a flashlight bulb in the shallow top of a tin baking powder can, and drove around in the dark basement of the duplex in which we then lived. Accidentally on purpose, I kept running into the steel posts that supported the floor above and broke light bulbs. I tried to make a pushbutton to operate an electric bell, but the complexities of construction eluded me. Finally, I made a switch of a coiled spring of wire, pulled by a cord and the bell rang when I pulled.

When I was very young, the only companions of my age were the children of relatives and friends of my parents, and children of neighbors. During my school years, I made some friends, but they were few in number through all my days of learning. My friends and I played with steam engines, electric motors and other gadgets as if these were manifestations of natural magic (as I suppose they are), to be invoked by spells that we might learn some day.

When I became an eager reader, my first favorites were the Oz books — one a year as a present at Christmas. These were full of plausible wonder. The author, Frank L. Baum, wrote that he was in touch with the magical land of Oz by wireless. I didn't believe this, nor did he intend that I should. I did believe, with many other bemused creatures, that all words must denote something. When I went on to the inferior Tom Swift stories, and later to science fiction, I did not much distinguish words from things, or science from magic. I had no help from the authors, who were not in a position to make such distinctions. Magic, science were arts by which men worked wonders. Words were as real as life.

For a good many years my romantic yearnings after the magic of science had no substance or understanding. As I grew older I went beyond words, but then I gave my heart to gadgets. I took photographs simply to be taking them. I built crystal radio sets, and later vacuum tube receivers, but I valued them more for their complexity than for how well they worked.

What part did the schools of St. Paul play in such interests? A friend and I wondered what we learned in grade school. Mostly, we couldn't remember.

About 1924 I left St. Paul, its schools, its state fair, and its lakes, for my parents moved to Mason City, Iowa. There my father, now in partnership with his brother, managed the millinery and ready-to-wear departments in Shipley's department store. Now we went to Clear Lake in the summer, instead of to White Bear or Bald Eagle. In Mason City I attended eighth grade and the first two years of high school. These were agreeable but not very strict. When I went back to St. Paul as a junior, I could not meet a minimum requirement in spelling and punctuation. I had to take a no-credit course, which has been of value to me to this very day.

In my junior year I had a course in physics. I dogged the steps of the physics teacher while he walked home, but I was deaf to what he was trying to convey in class and laboratory. I may have learned a few facts, but the point of the experiments escaped me entirely. To me laboratory work was playing with toys. That I was supposed to be going through steps that illustrated the discovery or verification of natural laws missed me completely. I had no idea of a process of discovery. Someone just knew.

Mathematics came easier. I could see some reason in it. Geometry appealed to me particularly, because I could see through, could hold in my mind, the whole process of reasoning, from assumptions to conclusion. Algebra was easy, too. I also had a good time in chemistry during my senior year at high school in

Long Beach, California. In some degree, I grasped the sense of laws of volume, pressure and temperature, and some basic ideas of elements and compounds.

All that I learned at high school was strangely disassociated from my continual tinkering. I built more and more elaborate radios, but I never became an amateur. Learning the Morse code seemed too difficult. I made a Tesla coil, driven by a Ford spark coil, and drew harmless sparks, but I didn't think through the operation of the device. I read a science-fiction story whose hero had a magical electronic belt that worked all sorts of wonders. I set out to build such a belt, but all that I could think of putting in mine was an unsatisfactory crystal radio receiver.

When I look back on my high-school years, I see a glimmer of the dawning of the idea that things can be understood, and that learning, in science at least, is understanding. But largely, in my mind the glamour of science was associated with Hugo Gernsback's popular technical magazines and with science fiction. It was because I read these that I regarded myself as a scientist. I remember the laughter in a high-school class when I asserted that "other scientists" also believed that there was life on Mars.

CalTech

Neither of my parents had gone to college, but neither of them, nor I, had any doubt that I would, or that I would pursue some sort of technical career. Where I would go and what I would study lay in deep doubt. None of us knew anything about the academic world.

My parents were far from wealthy. They moved from Long Beach to Pasadena, so that I could live at home. Caltech had a high reputation. Robert Andrews Millikan, of cosmic ray fame, had gone there in 1921 and created a first-rate school of science and engineering from what had once been the Throop Institute of Technology. In those days, Millikan stood for science in the public eye, as no contemporary scientist does.

At Caltech the freshman studies were the same for all students. I had to have some sort of objective. I decided that I would be a chemist. Freshman chemistry cured me of this aspiration. Lectures on the law of mass action baffled me completely. My laboratory work was a disaster.

From chemistry I turned my aspirations to aeronautical engineering, because I was deeply engrossed with gliders. But, I had difficulties with drafting. The quality of my lettering varied so erratically from day to day that the instructor asked if I had personal problems. I was bored and exasperated by drawing endless rivets in steel beams. If this was necessary in aeronautical engineering, I wanted none of it. At the time of my exasperation with drafting, my interest in gliding was waning.

I turned to electrical engineering, which seemed as far as possible from chemistry and beams with rivets in them. Too, it fitted in with my building and tinkering with radio receivers. This proved to be as happy a choice as if I had made it on rational grounds.

As an undergraduate, was I drawn to the Humanities faculty because they were there to teach, to serve the students rather than to do research? Or, was it because I was an omnivorous reader with a desire to write?

I received my B.S. in 1933, in the midst of the depression, and I couldn't get a job. Perhaps I didn't try very hard, and, indeed, I didn't know how to try. My indulgent parents supported me through three years of graduate work. This was a very good thing in most ways, though it delayed my facing and coping with the outside world.

I studied harder and with more interest during my graduate years. To get a Ph. D. I had to do research and write a thesis. What I did find interesting was to make a sampling oscillograph to record the waveform of a periodic wave. Many years later, this has become an important way of depicting waves of very high frequency. When the machine was completed, I wrote a thesis describing it.

Though the oscillograph I built worked, my thesis had its weakness. While I made some analyses relevant to the accuracy of the reproduction of the waveform, I did not verify these experimentally. I was still essentially a tinkerer, with some relevant skills in analysis.

Nonetheless, the thesis was accepted, and I passed an oral examination with enough credit to receive my degree Magna Cum Laude which was better than Cum Laude but not as good as Summa Cum Laude.

I haven't mentioned one of my teachers. In a course on electronics, S. S. Mackeown taught me what little I learned about the nature of vacuum tubes. He must have thought well of me, for I am sure that it was through him that I was offered a job at Bell Laboratories when I received my Ph. D.

Getting Started at Bell Labs

The date of my starting work at Bell Laboratories was to be September 1936. My parents, through a sort of general good will, and perhaps in celebration of my receiving my doctorate and getting a job, sent me on a trip through Europe during the summer. This was a slow journey by today's standards, by train and small steamer, bicycling in England and traveling by rail in France, Italy and Germany. On my return, I went back to California to see my mother and father and to collect my few belongings, and then on to New York, where I was to work at 463 West Street, in Manhattan.

I received little guidance except that I was to do research on vacuum tubes. I was told of some work by Philo Farnsworth, and of a high-frequency oscillator with no hot cathode that he had made. I finally visited Farnsworth's laboratory, no doubt with others, and I took back with me one of his oscillator tubes, which I tested and came to understand. It wasn't of any practical value. More important, I saw the electron multipliers that he incorporated in his image dissector and in other devices. These seemed to me to be crude and unduly complex in construction. So did the magnetically focused electron multipliers that Myron Glass, a colleague at Bell Laboratories, had built. It was Myron Glass who said, truly, "Nature abhors a vacuum tube," a remark later attributed to me.

At this time I came to know Bill Shockley, a new employee who was making the rounds of various parts of the research department. His effect on vacuum-tube research was electrifying, in his work with others as well as with me. It was because of him that a practical means for designing electron multipliers was worked out. At this point I really don't remember whose idea it was; it may well have been Shockley's. Certainly, he had a lot to do with the construction.

In the course of my work on electron multipliers, I became concerned with the noise in the output of these devices. Shockley worked out a theory. I simplified it. We published a paper on this in the March, 1938 issue of the Proceedings of the IRE.

Microwave Tubes

My work on electron multipliers resulted in an excellent device for which there was no use in the Bell

System. Somehow, I became involved with microwaves, which seemed to be the coming thing for' broadband long-distance communication. Later this work was to be turned to wartime radar, but when I first visited Harald Friis' laboratory at Holmdel, New Jersey, the emphasis was entirely on communication. What was lacking was satisfactory microwave tubes.

I set out to design a low-voltage klystron amplifier. I realized that the bunching and hence the gain would be most effective if the beam was slowed down in the drift space between the input and output cavities, so I put a low-voltage grid between these.

The result was an amplifier with a gain of around 10 db. My boss, J. O. McNally, was impressed. By that time Bell Laboratories had become involved in wartime, or pre-wartime, radar work, and he wanted to try the tube as an amplifier at the front end of a radar receiver. Accidentally, I made the grid between the input and output cavities negative rather than positive, so that it reflected electrons back through the input cavity. The tube oscillated like mad, and one could get on unprecedented 100 milliwatts from the input cavity.

I don't claim to have invented the reflex klystron; I stumbled on it. The virtue of the reflex klystrons we produced at Bell Laboratories for radar use was that they were simple, low-voltage devices that worked well.

At my urging, Gerry Shepherd set out to design a really satisfactory metal klystron with an internal resonator. Gerry learned about the capabilities of the machines for fabricating metal tubes, and he sat down at the drafting board and designed the X Band oscillator himself. The tube, the 723A, was a tremendous success, and inexpensive to fabricate. It was used in all American X Band radars. It came to be known as the Pierce-Shepherd tube.

After the war, Gerry Shepherd and I published a long paper in the *Bell System Technical Journal*, telling all we knew and all we had done concerning reflex klystrons. Gerry went away to the University of Minnesota, where, after a distinguished career in electronics, he became a vice president. I turned to other matters.

Traveling-Wave Tubes

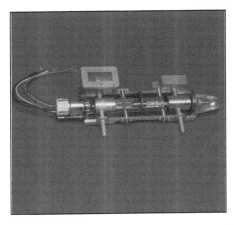

A traveling-wave tube developed at Bell Labs. (Reprinted with permission of Alcatel-Lucent USA Inc.)

Work on traveling-wave tubes occupied most of my time from 1945 until around 1955 — a period of thirteen years. Today, when vacuum tubes survive almost entirely as extremely high-power devices, and solid-state devices serve almost all low-power and moderate-power functions, the traveling-wave tube is still used in some communication satellites as a transmitter producing around 10 watts.

The inventor of the traveling-wave tube was Rudi (Rudolf) Kompfner. He was born in Austria, he immigrated to England, and on December 27, 1951 he came to Bell Laboratories. He was my very dear friend.

During this period I had made traveling-wave tubes, and they worked, giving gains of tens of db. This was at a time when Kompfner had become somewhat disillusioned with his invention, and worked for a while along other lines. He used to say that he had invented the traveling-wave tube, but that I had discovered it.

Certainly, I saw the value of the traveling-wave tube the very moment I learned of it. Kompfner had seen the tube chiefly as a low-noise amplifier. I saw it chiefly as a broadband amplifier. At that time I had visions of extremely broadband digital micro-wave signals, and the tube seemed the answer to my dreams. I was only partly right.

In late 1947 Bill Hebenstreit and I made an astonishing invention. This was the double-stream amplifier, a device in which two electron beams interact with one another to provide a wave of growing amplitude without any circuit. How wonderful to get away from the circuit! We thought that our idea would revolutionize microwave amplification. It didn't. The double-stream amplifier is a member of a large class of devices and inventions – wonderfully ingenious, and good for nothing.

I find that I continued to publish papers concerning microwave tubes as late as 1959, but my attention had turned elsewhere. In fact, I had always had wider interests, and by 1954 I was pretty tired of traveling-wave tubes. There was still a lot to be done, but it wasn't the sort of thing I could do, or would want to.

Communication

Within a few years of my going to Bell Laboratories, I developed serious interests in things other than vacuum tubes. I became fascinated by communication in a very broad sense.

For many years waveguides were to be the wave of the future, the right way to send great numbers telephone and television signals over land. This seemed obvious, yet optical fibers, cheaper and better, made waveguide systems obsolete before they came into use.

Shortly after World War II I met Claude Shannon, who had a profound influence on my life, through his creation of information theory, and in other ways. Eventually, I was to write a popular book on information theory, *Symbols, Signals and Noise,* published in 1961. Information theory gave prominence to digital communication, another recurrent theme in my life.

Chiefly, I think of PCM in connection with discussions of communication with Claude and Barney Oliver. I had known Barney briefly at Caltech, but I came to know him well while he was at Bell Labs. During our early association, before we moved to the new Murray Hill laboratories, Claude, Barney and I published a paper, *The Philosophy of PCM*, in the November 1948 issue of the *Proceedings of the IRE*.

I think that all of us expected PCM to be adopted quickly, but it wasn't. Eventually, transistors and cheap integrated circuits made PCM coding and decoding (analog-to-digital conversion, and digital-to-analog conversion) cheap and reliable. Optical fibers have made the broad bandwidths at which PCM is best cheap. The world of communication is becoming a digital world, and PCM the best method for transmitting continuous signals, such as voice.

Another result of my association with Claude was a brief and ineffective infatuation with artificial intelligence. Claude had built a maze-solving mouse, using relay circuits to provide the logical operations that enabled the mouse to find its way through a maze and to remember the path. Really, the mouse didn't remember anything. All the "intelligence" was located under the maze. A magnet under the aluminum floor of the maze pulled the mouse along, and circuits could detect when the mouse bumped into a barrier. Sometimes I have a wry feeling that this may be a pretty good picture of man's place in the world.

Whatever thinking machines and artificial intelligence may be or may not be good for, they weren't for me. Indeed, I've since become somewhat wary of artificial intelligence. The host of machines and systems that have transformed our life do things in ways that are different from, and better than, those of man. If, during our industrial progress, technology had provided us with mechanical men rather than indoor plumbing, vacuum cleaners, central heating, automobiles and so on, we'd have plentiful servants to carry the slops, chop the wood, sweep the floors, and haul us around in carts or sedan chairs, but this sort of life seems primitive to me.

In 1952 I was made Director of Electronic Research, reporting directly to Harald Friis. This gave me more incentive to associate with others in Friis's division. In 1958 Friis retired, and I became an Executive Director of Research under Bill (William O.) Baker, the vice president in charge of all research. Then and thereafter I had in my division work spanning a very broad range, at various times switching research, work on computers, radio research, waveguide research, optical transmission research, mathematics and statistics research, economics research, work on speech and hearing, and behavioral research, which included psychologists of various types.

I learned something of and thought about all of these fields, sometimes more, sometimes less. Particularly, I fell in love with speech, hearing, and all that has to do with sound, including musical sound.

Deming Lewis and I had another bee in our bonnets in those days. That was to record all business correspondence in digital form and send it from office to office over phone lines. Deming's people made primitive cassettes for recording the data, much like today's audio cassettes. They showed that such a system would work. I was full of enthusiasm, almost as much enthusiasm as I had at the same time for satellites. And, at that time I would have bet a good deal that this primitive first step toward the computerization of offices would become commercial before satellites would.

What a mistake! At the time I blamed others at Bell Laboratories and AT&T for not adequately pushing office-to-office transmission of correspondence and other data. In retrospect, I see that the technology of those days wasn't adequate. The mechanical keyboards available were awkward and costly. There weren't any integrated circuits, nor were there floppy discs. By contrast, the technology necessary for successful communication satellites was available, and satellites were pushed, in the Bell System and elsewhere.

While I was at Bell Labs I did one more piece of work in the digital field, this time in the switching of data. At that time, the Bell System was debating, endlessly it seemed to me, the establishment of some sort of digital communication network to provide the sort of data communication that the government Arpanet was, and still is, providing. One problem was the large initial cost of switching.

I had this in mind when an idea occurred to me, sometime around 1970, I think, of a data switching system in which the switching would be done by each user's equipment. Thus, the cost of switching would be small to start with and would grow as the number of users grew.

My idea was to use a "ring" system, in which transmission was over a closed loop or ring. A simple device generated empty digital packets of fixed length. These passed in turn through each user's terminal. If he wanted to send a packet of digits to someone else, he wrote the person's address and some message digits into the first empty packet that arrived; the packet now became non-empty and no one else could use it until the address and message were erased, either by the recipient, or by the device that generated the packets. My idea included ways to deal with non-delivery of addressed packets, and ways to interconnect various rings in a hierarchical system.

My interest in digital transmission continued through later years, especially during the two years

I spent at Caltech's Jet Propulsion Laboratory. And, the techniques of PCM are those used in the generation of musical sounds by computers, which I'll come to shortly.

Echo

The Echo 1 *passive communications satellite. (Reprinted with permission of Alcatel-Lucent USA Inc.)*

"BALLOON SATELLITE ORBITS; DELAYED MESSAGE HERALDS NEW COMMUNICATIONS ERA" said a page-one headline of the New York Times of August 13, 1960, the morning after the *Echo I* communication satellite was launched. In President Eisenhower's recorded message, transmitted from a Jet Propulsion Laboratory antenna at Goldstone California to a Bell Laboratories ground station at Crawford's Hill, New Jersey by reflection of a microwave signal from the 100-foot balloon satellite, he described the satellite as "one more significant step in the United States program of space research and exploration." It was, and I had a good deal to do with it.

I was asked to give a space talk to the Princeton section of the IRE (the Institute of Radio Engineers, now incorporated in the Institute of Electrical and Electronics Engineers, the IEEE). I had a great respect for the IRE, and I felt that a science-fiction type of talk would be inappropriate. What could I speak of? The idea of communication satellites came to me. I didn't think of this as my idea, it was just in the air. Somehow, I had missed Arthur Clarke's paper on the use, of manned synchronous satellites for communication.

In an article "Don't Write, Telegraph," published in *Astounding Science Fiction* in 1952, I had calculated the power necessary to transmit signals between the earth and the moon, planets and stars. It was easy for me to calculate the power needed to send signals from one place on the surface of the earth to another by means of a communication satellite. I made calculations for large balloon-type satellites that would merely scatter the radio waves they intercepted so that about a billionth of a billionth of the microwave signal transmitted would be picked up by an antenna on earth. I also made calculations for active satellites with radio receivers, amplifiers and transmitters, both for satellites at low altitudes, and for satellites some 20,000 miles above the earth's surface, the sort of synchronous satellites we now have, which hang over one point.

I was amazed and delighted at the outcome of my calculations. By using currently available microwave equipment, any of these sorts of satellites could be used to communicate across oceans. The power required on active satellites was very small.

My lecture was well received, and Professor Martin Summerfield of Princeton suggested that I write it up and publish it in Jet Propulsion, the journal of the American Rocket Society. This I did, and "Orbital Radio Relays" appeared in the April, 1984 issue.

Alas, there was then no way to launch any of the satellites I had considered. Also, I was somewhat concerned with the reliability in space of the vacuum tubes and the primitive transistors of that day. I talked with a few people at Bell Labs about communication satellites, but nothing came of it then. On October 4, 1957 the Soviet Union launched *Sputnik* and startled the world. For many, *Sputnik* heralded the realization of the old aspiration of science fiction — manned exploration of space. And, indeed, this old dream has been realized.

I was interested in communication satellites, but I couldn't put one up. That required persuasion. A dear friend of mine at Bell Laboratories, Rudi (Rudolf) Kompfner, the inventor of the traveling wave tube, became an advocate immediately. I set out to convince others.

The task was made easier by our finding that William J. O'Sullivan of NASA had built, for another purpose, just the sort of metallized balloon that I had written about as one possible communication satellite, a passive sphere that would reflect part of a signal back to earth. Indeed, O'Sullivan's balloon was of just the size I had considered – 100 feet in diameter. O'Sullivan had made the balloon because he wanted to measure the atmospheric density 1,000 miles above the earth through the drag on the balloon orbiting as a satellite. He hadn't been able to get NASA to launch it.

Rudi and I went to Langley Field and discussed the balloon, but not with O'Sullivan, whom I never met. We used an ohmmeter to measure the conductivity of the aluminum coating on the plastic balloon, and decided that it would reflect almost all of the microwaves that struck it. We took samples of the aluminum-coated Mylar back to Bell Labs, and this proved to be true. Rudi and I set out to get someone to launch O'Sullivan's balloon, both for his purpose and as a communication satellite.

Rudi and I wrote a paper that was published in the *Proceedings of the IRE* in March, 1959. This paper lauded all satellites. Particularly, it published earth-coverage diagrams for satellites in orbits lower than synchronous orbit. I think that it was just after the publication of this paper that we learned of the paper "Extra-Terrestrial Relays" that Arthur C. Clarke published in the Wireless World in October of 1945. In that paper, Clark had pointed out that when manned synchronous space stations had been launched, they could be used for worldwide communication by radio. Thereafter we cited Clarke's paper in everything we published.

In the summer of 1958, Rudi Kompfner participated in an Air Force summer study at Woods Hole in Massachusetts, a very pleasant place. He invited me up for a day, to talk about communication satellites, and particularly, about balloon satellites. The Air Force didn't help us, but the meeting was useful. William H. Pickering, the director of the Jet Propulsion Laboratory, which was soon to be transferred from the Army to NASA, was there. I had known him for many years. He was sympathetic with our idea, and agreed to supply a west-coast ground station if we could get the balloon launched.

This didn't mean that all difficulties were overcome. We had to convince the management of Bell Laboratories that launching an experimental communication satellite was a worthwhile idea. My boss, William O. Baker, the vice president in charge of research, seemed agreeable, and so did James B. Fisk, the executive vice president. But Mervin Kelly was the president of Bell Labs. He called a luncheon meeting, probably in the latter part of 1958. His attitude was very negative. He asked an able mathematician to make a system study, which took some time and turned out to be very negative. Really, Kelly decided against the *Echo* idea right at the meeting. Rudi Kompfner thought that *Echo* was dead, but I just didn't hear what Kelly said.

Kelly was a great hero of mine, and a great leader of Bell Laboratories. Even great men can be wrong. Fortunately for *Echo*, Kelly retired early in 1959, Fisk became president, and opposition within Bell Laboratories turned to support. Early in 1959 NASA and Bell Laboratories somehow reached an agreement concerning Project *Echo*. *Echo* was to be NASA's project, and NASA would supply the balloon and would launch it. Bell Laboratories would build an east-coast ground station and NASA would lease it. The Jet Propulsion Laboratory, now a part of NASA, would build a west-coast terminal. Indeed, there is still an *Echo* antenna at JPL's Goldstone site, built for Project *Echo* but used since for other purposes.

It had been one thing to propose satellite communication, to make calculations concerning the feasibility of microwave communication by means of satellites, to locate O'Sullivan's balloon, to stir up the enthusiasm of people at Bell Laboratories, JPL and NASA. It was something else to carry the project through. At Bell Laboratories we had a wonderful resource for making an east-coast ground terminal. That was the Holmdel Laboratory. Not the new, big one, but the wooden laboratory that Harald Friis had made into a world center of microwave research. Harald had retired in 1958, and Rudi Kompfner was now head of the laboratory. Harald had brought wonderful people to work there. Because of Rudi and me and the technical value and attractiveness of the Echo experiment, these people were enthusiastic and eager to go ahead. Indeed, some preparation had been made before the agreement with NASA. At the end of 1958 Kompfner had used money available in the budget to order a lot of aluminum for a receiving antenna.

A lot of time passed between the initiation of the *Echo* project at Bell Laboratories and the successful launch on August 12, 1960, and much was done.

Both Rudi Kompfner and I became spectators of the excellent progress of the Echo terminal at Crawford Hill. We were gratified when signals were bounced off the moon in November 1969. Signals were bounced off the small *Tiros* satellite on May 11 and July 29 1960. There were suborbital "Shotput" launches of *Echo* balloons on January 16 and February 27 of 1960, and signals were bounced off the balloons so launched. I was disappointed when the first attempted launch of *Echo* failed on Friday the 13th of May, 1960. Happily, the *Echo* satellite was launched successfully on August 12, 1960.

Both Rudi Kompfner and I were so far from participation in the technical details of the work that we were sent from the operating area of the terminal to another small building, equipped only with a loudspeaker over which we could hear any received signals. There we heard President Eisenhower's recorded message so clearly and noiselessly that we at first thought that it was a local transmission from Crawford Hill to the JPL station at Goldstone. Then a fault of pointing caused a momentary interruption of transmission, and we realized that we had been listening to a signals transmitted via satellite from Goldstone to Crawford Hill.

My dream had been realized.

Telstar

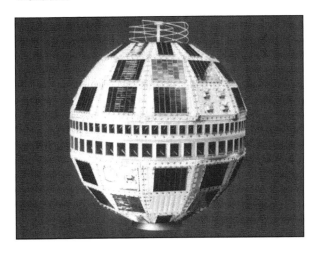

Telstar: The first active communications satellite. (Reprinted with permission of Alcatel-Lucent USA Inc.)

In the research department we had started work toward an active satellite at a modest level even before the launch of *Echo*. This work followed a satellite design set down by Roy Tillotson on August 24, 1959. The proposed satellite would have been at an altitude of 2,500 miles. It would have had an essentially omni-directional radiation pattern, and would have had a transmitter power of one watt. The plan was to use broadband frequency modulation, with a hundred-megahertz bandwidth. This would allow the transmission of one TV channel or of several-hundred telephone channels.

The research department of Bell Laboratories was small, around a tenth of the whole Laboratories, and those who could devote their time to satellite work were few. In the summer of 1960 the primary responsibility for an active satellite experiment was transferred to one of the vice-presidential areas of development, headed by Marcellus B. McDavitt. A. C. Dickieson was the executive director of the division responsible for the work, and Gene (E. F.) O'Neill was the project-engineer for what became *Telstar*.

Telstar involved problems of a scope and magnitude far beyond any we had faced in *Echo*. The transistor and the traveling-wave tube were key components, but they had to survive a rocket launch and survive for a long time in space. The space background that the Whippany laboratories had acquired through their military work, especially, work on the Nike missile, was invaluable to *Telstar*.

I am somewhat embarrassed to recount that Harold Rosen and his colleagues Don Williams and Tom Hudspeth visited the Holmdel Laboratory in

1960, when we had Tillotson's proposal for a lower orbit active satellite in mind. Rosen told us about his highly ingenious design for a synchronous satellite. He showed pictures of a large antenna built into the ground, aimed fixedly at the point in the sky over which his satellite would hover. This was irrelevant. The real problem was the difficulty and cost of the satellite, not the cost of a ground antenna, tracking or not tracking. I concluded that Harold Rosen was irresponsible, and I didn't pay enough attention to what he had to say. Actually, he was (and is) a tremendously ingenious, inventive, careful engineer. When he saw us, he was desperate to get his satellite built and launched, and he used every argument he could think of, good and bad.

Satellite communication via synchronous satellites is an amazing resource. It gives us visual access to all of the world. It links together the telephone networks of all nations, small and large. It is our best resource in mobile communication. But, it is not as powerful for communication within and between densely populated areas as are optical fibers. Further, two-way satellite traffic suffers some degradation through the time delay of almost a third of a second to the satellite and back, or almost two thirds of a second round trip.

Echo and *Telstar* established satellite communication firmly as a part of the future of telecommunications. Neither satellite could have been built and launched without NASA, or without the American missile programs (all the launch vehicles prior to the Space Shuttle were adaptations of ballistic missiles), or without the expertise and hard work of Bell Laboratories, of which I was one employee. What was my part in this?

Had I not been in my position at Bell Laboratories at that time, *Echo* would not have been launched, and *Telstar* would not have followed. I was fortunate to be the executive director of the part of the research department most crucial to these undertakings. Better yet, I was on good terms both with my bosses and with the people qualified to do the work. They respected my technical ability and were willing to pursue the opportunity that I pointed out.

The Human Equation

Around 1968, when I became an Executive Director, I acquired a group of behavioral scientists. When I found acousticians and psychologists in my division, I knew that I had much to learn. I'll discuss the behavioral science work only briefly.

I dealt with animal psychologists (only one), social psychologists (a few) and experimental psychologists (more). From the animal psychologist I learned much about the behavior of pigeons that fascinated me and all visitors. Interesting as this work was, its implications for the Bell System were obscure. The animal psychologist must have sensed this, for he left and became a university professor. The social psychologists seemed to be working on things of great potential importance to the Bell System. Alas, after I had made every effort to involve them, including a deluxe tour of a telephone operating company, the social psychologists drifted away.

This brings us to speech and hearing, with which I was concerned much more closely than with the psychologists. The man who was put directly in charge of this work was Ed David (Edward E. David, Jr.), who later became an executive Director at Bell Labs, then President Nixon's Science Adviser, and, after holding other positions, retired from the presidency of Exxon Research and Development to pursue other interests.

Ed and I both gave a lot of thought to the vocoder, an elaborate speech-encoding system invented many years before by Homer Dudley. We had ideas that were very hard to try out, for it took a long time to build the complex circuits. This problem was solved, not by me, through computer simulation.

This computer simulation work revolutionized research on all speech-processing devices and systems. Computer simulation made it feasible to attack old problems with much more powerful resources. One of these old problems was speech recognition — the "voice typewriter", if you will. A lot of work was done on speech recognition while I was at Bell Labs, and a lot more has been done since. I watched the work with interest. It seemed always to be promising, never to succeed in doing anything useful. I got a little fed up, and in 1969 I published a letter to the editor of the *Journal of the Acoustical Society of America*. The title of the letter was "Whither Speech Recognition?"

Work on computer simulation of sound-processing systems led to something quite different that was to become very important in my life. The original idea of computer simulation was to start with samples of a sound wave and use the computer to produce samples of the sound wave after it had been processed by a complex system, analog or digital. But — it occurred to someone — why not use the computer to generate a sound wave in accord with prescribed rules?

A great deal of effort went into generating samples representing speech. At Bell Labs, Max Mathews, John Kelly, Peter Denes, Jim Flanagan. and others were pioneers in such work. As a part of this, work, a computer sang "A Bicycle Built for Two" ("Daisy, Daisy, give me your answer true..."). Arthur Clarke heard this, and the computer Hal sang the song in the movie 2001. Today computers readily talk to their users, usually with an electronic accent.

Speech is not the only organized sound important to man. Music is, too, and computers can be used to generate musical sounds. As far as I know, this was first seriously undertaken by Max Mathews around 1957, at the request of Newman Guttman. Guttman, a phonetician, had an idea for an ingeniously organized sequence of vowel sounds and other sounds, a special sort of *avant-garde* musical composition. Max used the computer to generate this specified sound sequence. To me, it sounded awful.

Was this the fault of the computer, or of Newman Guttman? I asked Max to play on the computer a little piece that I had written for recorder and piano, and to use simple but plausible waveforms. To me, this sounded much better. I was hooked, and so was Max. Over succeeding years, Max wrote sound-generating programs of increasing complexity, culminating in Music V. Music IV had been written in the assembly language of the IBM 7094, and when that machine went away, Music IV became inoperable. Max commemorated its demise in a composition called Swan Song, the last sounds ever generated using Music IV.

For a number of years, Max and I composed little pieces for the computer, and induced others to do so. Some of these were recorded through the enterprise of Bruce Strasser of Bell Labs' public relations department on a disk titled *Music from Mathematics*. We sent copies to various musicians. Through Max's enterprise, he and I have pieces on the Decca records Music from Mathematics and The Voice of the Computer, both out of print. Though we used the computer to play our pieces, we were anxious to interest talented composers. In our effort to make something of computer music, Max and I invited young composers to Bell Laboratories, sometimes for short periods.

Earlier, I recounted instances of valuable advances that were not planned. Rather, they arose in the course of sensible work, but were good for something other than their original purpose. The Pierce gun and the traveling-wave tube are examples. So, I think, is the Pierce ring. Some wise man said "It is very hard to predict things, especially the future." Planning is fine in constructing new examples of things that have already been built, as, bridges, office buildings and sewage treatment plants. But, too much planning constrains rather than facilitates progress.

Diversions

Some engineers and scientists become proficient in serving their profession, and mankind, through work for technical societies and for the government. Bill Baker is one of these. To me, such activities have seemed diversions from my technical work.

For engineers, the best sort of publication is patents, for these establish precedence over the work of others. While at Bell Labs, around eighty patents were granted in my name. I have a list of the obscure titles that don't remind me of what the inventions were. Patents are written by patent attorneys.

My technical papers and books I wrote myself, and I enjoyed writing them. I read science fiction avidly and uncritically. It had an effect on my aspirations that was perhaps similar to that of stories of saints and miracles.

When I was at Caltech I was much interested in writing. As I wanted to be published, it seemed natural to me to write science fiction stories. My first was published in 1930. I have had twenty-two short stories published in all, the last in 1973. Two have been included in best-of-the-year anthologies. I got to know most of the leading science fiction writers of the day. I did not become a leading writer, though some of my early stories are occasionally republished in anthologies. I published them under the pseudonym J. J. Coupling in order to avoid the Bell Laboratories release procedures for technical articles, procedures that had nothing to do with the sort of things I wrote.

Bill Baker, whose service in Washington is no diversion, but a chief feature of his life, projected me into a different world. I found myself on his advisory committee to the NSA (National Security Agency). I found NSA to be highly competent and to do excellent technical work. This I attribute partly to the fact that it has a continuing task; it must produce results, and that security in some measure protected it from meddling by Congress. I found later that CIA technical activities were also good, for the same reasons. Some other parts of CIA seemed less brilliant. I hope that the technical work of NSA and CIA are still as good as I knew them to be

I am sure that it was through Bill Baker that I came to be a member of PSAC (President's Science Advisory Committee) from 1963 through 1966.

Toward the end of my term on PSAC I realized that I hadn't contributed much, and I organized a panel on computers in higher education. *Computers in Higher Education* was issued as a Report of the President's Science Advisory Committee in February, 1967. It would probably have done more good had I pursued the matter in Washington after the report had been issued, but I am no Bill Baker.

Perhaps as a result of my service on PSAC, I was asked to be chairman of a committee of the National Academy of Sciences National Research Council that was formed to look into machine translation. The thorough and sensible report, Language and Machines, Computers in Translation and Linguistics was published in 1966. I doubt if it had much effect in a giddy world. Computers have an important role in translation, especially in text editing and in the production of specialized dictionaries tailored to various fields and to particular documents. Computers have found some use in translation itself. However, there was, and is, and will continue to be a lot of expensive nonsense in machine translation.

People and Places

People are the most important part of our lives and our work. What we do is done for ourselves and others. What is done is done by individuals, never working in complete isolation. People are inseparable from places. What people do, what they think and feel, are influenced strongly by where they work.

In 1971, four years short of retirement at Bell Labs, I was made a flattering offer as professor of engineering at Caltech, and I accepted. While I was at Caltech, I had a number of very able graduate students who wrote theses on diverse topics. After I was made emeritus at Caltech in 1980, I served part time at JPL for two years as Chief Technologist. This was a staff position directly under the director, Bruce Murray. I came to understand JPL and its people, and to admire them a great deal. Since 1983 I have been associated with Stanford. I work at CCRMA (Center for Computer Research in Music and Acoustics).

For most of my career, I worked in a laboratory, Bell Labs, rather than a university. JPL, though managed by a university, is another laboratory, and I came to know something about other industrial and government laboratories.

I don't think that it is nearsightedness that makes me say that the Bell Labs, where I worked for thirty-five years, was the best industrial laboratory in the world, and perhaps the best laboratory in the world. Bell Labs had a broad but clear purpose, it was responsible for advances in the American telecommunication system.

Some industrial laboratories are either neglected or meddled in by the organization to which they belong. AT&T did not meddle in Bell Labs work, and especially, not in research. Bell Labs was the goose that laid the golden eggs. But in part, Bell Labs was protected by strong leadership, and this helped Bell Labs to produce good and appropriate things.

There was good leadership throughout Bell Labs; Harald Friis was an example of this. There was good leadership at the very top, too. Mervin Kelly was a great leader. When he was president of Bell Labs he attended internal technical meetings. That he understood what was going on you could tell from his sharp comments. His views on Labs objectives and work were his own; they weren't a consensus or selection of what others down the line told him.

Jim (James B.) Fisk succeeded Kelly as president of Bell Labs. Fisk was a highly intelligent man, less forceful than Kelly, very pleasant and friendly. Bill (William O.) Baker followed Fisk. I had already left Bell Labs when he became president; I had known Bill when he was my boss and vice president in charge of research. Bill knew a lot about the people in his area. I found at merit review and salary review meetings that he knew more about some of my people than I did.

Since divestiture, the abolition of the Bell System, many people have asked me what effect this has had on Bell Labs. I think of some things that Ralph Bown said when he was Director of Research. Bown posed a rhetorical question. What is the Bell Laboratories? he asked. If we marched all the people out and destroyed the buildings and the equipment and the records, would

Bell Laboratories be destroyed? No. But, if we preserved the buildings, equipment and records and removed the people, Bell Laboratories would be destroyed.

The facilities of Bell Labs have not disappeared. Many of the people are still there, especially in the research department. What has been lost is the mission of providing the future of American telecommunication. When this purpose vanished the Bell Laboratories that I knew ceased to be.

Bell Laboratories is a suitable subject for a historian. Its beginning can be traced; the end of its original mission and of the beginning of its transformation into something different was the date of divestiture.

MANFRED R. SCHROEDER

Manfred Schroeder. (Reprinted with permission of the University of Göttingen.)

A Dream Come True

Bell Telephone Laboratories in Murray Hill, New Jersey was, of course, a dream place to start one's career as a mathematically inclined physicist. I am not implying that people at other times and other places cannot enjoy their work as much as we did. But there is a general perception that Bell Labs was something special — a "national resource." This did not come about by accident: Bell management decided early on that freedom to pursue one's own ideas and stable, long-term funding were the best well-springs of innovation. Some of the greatest advances in the past have come from taking the long-range view.

How did I get to Bell Labs from distant Germany in the first place more than fifty years ago? Let me explain: I was studying at the University of Göttingen when a noted linguist from Bonn, Werner Meyer-Eppler, visited us around Christmas in 1951 to give a talk at the General Physics Colloquium. I don't remember the details, but he talked about Shannon and Information Theory. Perhaps this was the first time I had heard the name Shannon and I was immediately fascinated. This sounded much more interesting than the nuts-and-bolts of experimental physics, but, during the Christmas party after the talk, people professed not to have "understood a word" of the talk. The chairman of the colloquium, Richard Becker, even seemed a little miffed when I dared pronounce the lecture "interesting." How dare I! How could I find a field that was devoid of *relativity* and the *quantum* to be interesting? Undeterred, I approached my professor, Erwin Meyer, concerning whether he could recommend me for a job at Bell Labs. Unfortunately, the answer was "no," because he remembered one of his students before the war (at the height of the Jew-baiting pogroms in Germany) had wanted to emigrate but at the time Bell said they didn't hire any foreigners. End of dream to join Bell! Or so it seemed.

Two years later I learned that William Shockley, co-inventor of the transistor, was coming to Göttingen actually *looking* for bright students for Bell. Well, I flew back to my professor telling him, "they *do* take foreigners…they are even *looking* for them." Fine he said, I am corresponding with one of the research directors of Bell and I will put in a few lines for you.

Two weeks later, I received an invitation for an employment interview by said research director. The interview apparently went well, and after six weeks I received an offer of permanent employment by Bell Labs. Of course, I accepted the Bell offer — in fact, I would have worked at Bell for nothing, but I didn't mind getting paid for what was for me a dream come true. So on September 30, 1954, I arrived in New York on the Italian liner *Andrea Doria* — still afloat then. My future supervisor, Ralph LaRue Miller, and the director, Winston Kock, met me at the pier.

At the Labs, Win Kock encouraged me to continue my thesis work on concert hall acoustics, but I figured — this being the telephone company — I better do something more germane to the telephone business. So I picked "speech" as my new research field — and speech it was for the next fifty years. Nobody objected to my total ignorance of this field. Electronic engineering, especially, was an enigma to me, but I learned a lot by "osmosis," you might say. Bell Labs was an ideal place not only for doing research but also for learning. The doors were wide open and the people friendly and communicative. Nobody was guarding any "trade secrets." Not getting any monetary rewards for our inventions helped, of course, to foster this open atmosphere. I, for one, was quite happy with the $1 I received on my first day of work for all my future inventions.

Just one tiny but typical example: one day I was reading a technical paper bemoaning "the metallic twang" of existing artificial reverberators for electronic music. I thought what these audio types needed was a reverberator that, when seen as an electrical filter, has an all-pass frequency response. So I swivelled around in my chair and asked my office-mate, electrical engineer and country musician Ben ("Tex") Logan, "are there any all-pass filters with an exponentially decaying impulse response?" Ben's answer was "yes," and *colorless artificial reverberation* was born. As far as I know, Bell Labs didn't make a penny on the patents. But you can now find colorless reverberators and artificial stereo (now called soundscape, surround sound, or virtual acoustic images) in practically all electronic music instruments, Yamaha and the rest.

Growing Up in Hitler's Germany

What of my childhood? I was born in the house of my grandparents, a large director's villa near the entrance to the coal mine *Gewerkschaft Westfalen* where my father worked as an engineer and my grandfather, Robert Kraemer, was director of below-ground operations. This was in a city called Ahlen in what is now the state of North Rhine-Westphalia. *Westfalen* was the northeastern-most mine of the Ruhr coal fields and they had to go to a depth of 1 km and more to find the black gold.

Not far from the house was a city bus stop where I befriended one of the drivers, Karl Jaxtin, who took me on extended bus tours around the city. This was before 1933, but Jaxtin was already a Nazi, an *S.A.-Mann*, feared by the local communists among the miners for his combativeness. During the war he became a tank driver and perished in the early part of the attack on the Soviet Union.

My first "school" was a Roman Catholic kindergarten, St. Joseph's. I remember my nickname from those pre-school days: *Kohlenklau* (coal thief). I don't know how I acquired that moniker; I certainly never stole any coal as a child and I wasn't particularly dirty. But most of my friends were miners' children and, during the depression, *Kohlenklau* was a meaningful concept among them. After kindergarten, I attended the parochial *Volksschule* (grade school) for four years.

One thing I remember from my early school days was in 1934 when I saw police patrols near our house: the *Röhm Putsch* was in full swing with many "disloyal" *S.A.* men murdered by Hitler and his gang — Goering, Himmler, Heydrich and members of the loyal, elite *Schutzstaffel* (S.S.) — during and after the "Night of the Long Knives," as it became known. Although barely eight years old, I sensed that something was very wrong with Germany's leadership. But we, as children, still got a kick out of the rambling rantings of the *Führer*, sometimes lasting three hours or more.

At age ten, I left the *Volksschule* and transferred to the local *Gymnasium* (high school emphasizing the humanities, classics, Latin etc.). In the meantime my father, a World War I combat pilot, had joined the reconstituted German *Luftwaffe* and, after a brief stay in Göttingen, was assigned to the gigantic *Luftzeugamt* (air force supply center) Rotenburg on the Wümme, halfway between Bremen and Hamburg. So, in early 1937, the family moved to Rotenburg where I started to attend the Middle School.

In 1937, after I turned ten, I had to join the *Jungvolk*, the junior branch of the *Hitlerjugend*. I had volunteered when I was seven or eight because I was envious of the boys marching in the street outside our house. This looked like great fun. I applied and was accepted. But, being the smallest kid, I always marched in the last row with "giant" strides and was still unable to keep up with the older boys. My uniform was, of course, much too large for me and my belt was sometimes slipping down near my knees. Everyone, my parents included, got a great laugh out of watching little Manfred parade by the house. Later, when I *had* to join, it was no longer such fun. I remember with particular dread the *Kriegsspiele* (war games) where we had to box and wrangle with each other.

After two years at Rotenburg I transferred to the *Lettow-Vorbeck* high school in Bremen named after the "heroic" World War I defender of German East Africa (now Kenya). Since we continued to live in Rotenburg, I became a *Fahrschüler* (commuting student), always leaving Bremen at 13:26 for the forty-five minute return trip home. Fifty years later, the train is still running, but it takes only twenty-five minutes now.

Around the same time (1938) also came my only chance to see Hitler in the flesh. My father and I had travelled to Hamburg (by train, I believe) to witness the launching of the KdF cruise ship *Wilhelm Gust-loff* (or perhaps her sister ship the *Robert Ley*) at the Blohm & Voss shipyard. KdF stands for *Kraft durch Freude* ("strength through joy"), the Nazi attempt to assuage the German worker after the rape of their unions on May 2, 1933. One thing I remember from the *Gustloff* launching was the few and isolated shouts of *Heil* when the *Führer* appeared on the platform. I was amazed because on the radio we always heard the endless *Heils* when Hitler loomed. Were we being hoodwinked? Not necessarily, because, it was explained to me, the Blohm & Voss workers were mostly former communists.

Soon after the war started, British bombers appeared over northern Germany and my parents decided to move to East Germany, to Liegnitz in Silesia, where my father had just been transferred. They didn't want to get separated again — and they were afraid of bombs. Of course, the real bombing campaign didn't start until several years later but my parents never regretted their decision.

I saw the downside of the *Führer's* policies first-hand when a distant relative, a customs official, showed me around the Bremen docks and warehouses where I caught sight of a lot of furniture in storage. "Belongings of Jewish emigrants," my guide explained "but they cannot be shipped on during the blockade of the German ports during the war". The loss of the Jewish citizens after 1933 created a void in German intellectual life that — seventy years later — has still not been filled, I believe. Without that deprivation, Germany would be a much more vibrant country than it actually is.

To avoid constant school changes, my parents sent me to a boarding school in Ballenstedt in the Harz Mountains: modern buildings surrounded by forests and a beautiful little lake nearby on whose shore I spent many a dreamy moment. This was in September 1941. I had just turned fifteen, feeling very lonely, far from family and old friends.

Like most, if not all, boarding schools in Germany during the Nazi period, Ballenstedt was infested by ideology. On my first day, my homeroom teacher asked me whether I went to church regularly and when I answered "yes," he said that I would be free continuing to do so, but I would be the only one in the entire school. During the summer of 1942, our class was shipped to German-occupied Poland, to a village then called Nussdorf to help with the harvest. I was assigned to a farmer from one of the Baltic states, a dour, ethnic German who was given a former Polish farm to run which, for all I knew, may have belonged to a German before 1919. We had to get up at 5:00 in the morning to harvest buckwheat overgrown by weeds, especially thistle. It was very painful work and as the sun rose during the day it became hot as a furnace. One of the few good things I remember was the creamy milk from "our" cows.

War Duty

In mid-February 1943, around the time of Goebbel's infamous "total war" speech, I was drafted — barely sixteen years old — to serve, as an air force auxiliary (*Flakhelfer*), with an anti-aircraft battery near Stettin on the Baltic Sea. During unpacking, the battery commander noticed a book entitled *Funkmesstechnik* (radio measuring techniques). But *Funkmess* was also the German word for radar, so I was assigned forthwith to the fire-control radar of our battery, comprising six of the renowned 88-mm guns. There were a total of eighteen batteries guarding a huge "hydrogenation" plant that converted coal into gasoline. Once or twice a week there was an alarm: approaching enemy aircraft. But they never got very close and, after an hour or so, we could go back to our bunks. This got to the point that I never even got up during alarms.

But one night, April 21-22, 1943, matters got serious. I pulled my pants over my pyjama and wrapped myself in a greatcoat. By the time I reached my position at the radar, my screen was filled with a forest of "blips," barely 10 km away. How to pick *one* target in this mess? I selected one of the bigger blips more or less at random. But by the time my two buddies (responsible for range and elevation) had zeroed in on the selected target, it had disappeared and I had to select a new one. The same scenario repeated itself over and over again during the entire night. Where did the vanishing targets go? They were not shot down; they disappeared even before our guns could fire a single round.

Pretty soon it dawned on me that the big blips on my radar screen were not really individual planes but the result of random interference of the radar reflections from *many* planes that would add up or cancel each other depending on the (slightly changing) distances between the planes. Thus, we were shooting at "thin air," random blips, all night. So we moved our servos as we had learned from our single-plane exercises so at least the (analog) fire-control computer wouldn't go haywire with random input data. Almost needless to say, we didn't shoot down a single plane all night.

In early 1943, an Allied plane was downed near Rotterdam without the (navigational) radar self-destructing so that it fell into German hands nearly intact. Hermann Göring was taken aback when he saw a precise map of Berlin, including its lakes and rivers, on the scope of the reconstituted enemy radar. Göring then issued an order, in his capacity as president of the *Reichsforschungsrat* (Reich Research Council), for all young Germans with a background in electronics to volunteer for special radar training to catch up with the Allies. I, together with some 400 other 17-year olds, followed the call and reported for duty at the camp "Prinz Eugen" in the *Westerwald* on Saturday, October 23, 1943. In the next six months, we learned all about Maxwell's laws, German radars and radar receivers (preferred on U-boats), and Allied radars as far as was known.

After the *Westerwald*, my next military assignments were with the Navy: first boot camp and then the Radar School on the Baltic island of Fehmarn and the Radar Observation Station in St. Peter-Ording on the North Sea. In Fehmarn we learned about German countermeasures against the tinfoil strips ("chaff") that had blinded the German radars ever since the attacks on Hamburg in July 1943. The new German radars exploited the Doppler effect to distinguish between slow-moving strips and fast airplanes.

At the end of the war, I found myself with coastal radar in Holland — huge installations that could pick up a plane over distant London once it was 200 meters above ground. The Dutch navy officers, who had missed radar while in hiding during the war, were impressed and wanted us to repair them for their use in coastal shipping. But the Allies insisted on our dismantling them, being afraid (in the summer of 1945) that the Soviet Army might advance to the coast opposite Britain and use them against their former allies.

Of course, everybody was happy that the war was over. The only reason we youngsters were sorry that Germany had actually lost the war was our fear that there wouldn't be any more nice German *Tanzmusik* on the radio. But it didn't take us long to discover that the Allies were broadcasting even nicer jazz ballads and swing music. Another circumstance that lightened our burden of having lost the war was that we were sitting on endless supplies of molasses (in oil barrels!). We soon discovered how to ferment the stuff, distill it, and collect it in a beaker, drop by precious drop.

Incidentally, at the end of the hostilities in May 1945, we were classified as "capitulated personnel" — outside the Geneva Conventions. But in our case, in Holland, it meant Canadian army rations and even home leave, which allowed me to ascertain that my parents and sisters had survived the war unharmed and our house in heavily-bombed Hamm was still standing. Indeed, as I approached our house, after having traversed plains of rubble, I didn't dare ring the doorbell. Who would answer? Were my parents still alive and had they managed to flee from distant Silesia? So I walked around the house and saw, through a basement window, a large store of apples that maybe only my mother could have scrounged up in those desperate days. I finally entered the house through a back door into the kitchen. And I didn't have to ask who was still alive. They were all there in that one room: my parents, my sisters, and even my grandmother.

In May 1947, I was discharged from the German navy, receiving one pair of long johns and a very nice blanket for my services to the Third Reich. (The Nazis had, of course, promised everyone a farm in Kazakhstan, but I was never very fond of farming anyhow and the blanket came in very handy: my mother had it dyed a fashionable dark blue and turned it into a pretty overcoat for my older sister.) After my discharge, in July 1947, I travelled to Göttingen to apply for admission to the university, still one of the most prestigious in Europe.

Off to the University of Göttingen

After having been released from my captivity "outside the Geneva Conventions," I had to face the question of what to study and where. I was always strongly attracted to mathematics but my grandmother thought I should become a physicist. Fine, but *where* to study physics in those days? In Göttingen, of course, where several great physicists were in residence: Nobelists Max Planck, Max von Laue, and Werner Heisenberg, whose quantum mechanics we youngsters were all eager to understand.

But quantum mechanics we *didn't* understand and so I attended, in one of my first semesters, a course by Heisenberg on the shell model of the atomic nucleus — and didn't understand a word: Spin-orbit coupling! What is *spin-orbit coupling*? After a few lectures I dropped out and decided to learn some basic physics first.

Later I met Heisenberg on a more personal basis. I was asked to guard his house on Merkelstraße and his in-laws (the Schumachers) while the family were taking their annual vacation in the Alps. The house had been broken into on more than one occasion and valuable musical instruments had been stolen. Another reason for my moving in with the Heisenbergs was that around 1946 suspicious characters had been observed casing the house. The border with the Soviet zone was only 20 km away and rather penetrable immediately after the war. The fear was that these guys were Russian agents preparing to capture Heisenberg and exploit his knowledge for the Soviet atomic bomb project. I never believed this cloak-and-dagger story but his son Martin, a biology professor in Würzburg, in a recent conversation, confirmed the story, except that the suspicious characters were from a *Western* secret service.

With the *Vordiplom* in hand, I registered for the obligatory advanced lab course in physics, which didn't really stoke my interest — except when I had a chance to *umfunktionieren* (turn upside down) one of the experiments. For example, we had to measure the speed of light on a long cable (curled up in the basement of the physics building). The idea was to send a short electrical pulse through the cable and measure the delay with which it emerged at the other end. I thought, if I just connected the two ends of the cable through a feedback amplifier and cranked up the gain, the cable would start "singing" at a frequency that is simply the reciprocal of the sought-after delay. This new approach to measuring the speed of light on a cable, which I considered almost self-evident, was rated a "major breakthrough." The next day I was offered a position at the so-called "Third Physical Institute" at the University in order to conduct research for a master's thesis. Thus, while I had started out in theory, I was now an experimental physicist at a renowned institute that, four years later, smoothed my path to the U.S. and the prestigious Bell Laboratories.

After finishing my work for the degree of *Diplom-physiker* (comparable to an M.S.) in just four years after coming to Göttingen, I had enough of university life and went into industry: Grundig Radio. In early 1952, they were just beginning to set up production of (black-and-white) TV sets. For their production line, they needed various testing devices and I joined that effort building a square-wave generator that could switch between its two states ("black" and "white") in just a few *nano*seconds — an unheard-of speed in the early TV manufacturing days. When I told the Grundig managers that I had promised my Göttingen professor to return after six months in industry, they offered to increase my salary from 500 to 750 Deutschmarks and then even hinted I should name my price. But I made the right decision by honoring my word to Professor Erwin Meyer and starting work on my Ph.D.

I had always felt a great attraction to statistics and when I plunged my microwave antenna into a large discarded U.S. Army biscuit tin (after the last crumbs had been shaken out), I saw on my scope thousands of resonances, a veritable "forest" of spikes, all of different height and seemingly randomly distributed in frequency (around 10 Gigahertz). This looked interesting, to say the least! But Meyer wanted me to do *acoustics* (his field) for my thesis. I didn't want to part with my microwave statistics, however. So we compromised: I would investigate the random resonances of microwave cavities and he would interpret the results in terms of concert hall resonances. Thus, my published thesis bore the unlikely title "The distribution of resonances in large rooms: Experiments with microwaves."

I started working on my thesis in October of 1952 and was able to hand it in before the end of the following year — after just fifteen months, which must have been a record for a Ph.D. thesis in experimental physics. On February 26, 1953 I passed the orals for my degree (*Dr.rer.nat.*). Seven months later I was on my way to the U.S. to become a Member of the Technical Staff of what was then called Bell Telephone Laboratories.

My Early Days in the U.S.

Besides having to adjust to new technical fields, the social adjustments also were not always easy. I was twenty-eight when I joined Bell and still unmarried. This was somewhat unusual at the time and my Executive Director, Bill Doherty, introduced me to one and all with "This is Dr. Schroeder. He just joined us from Germany—and he's a *bachelor*." Or, as my old math professor from Göttingen, Wilhelm Magnus, then at New York University, put it: "This is not a country for bachelors. Either you will return to Europe within the year or you will get married." Well, he was right; I *did* get married, just seventeen months after my arrival in the US.

Another thing that surprised me was that people were not only very accessible but — I thought —

quite humble, like S.O. Rice of mathematical noise fame, one of my heroes even before joining Bell Labs. His office was at the old New York headquarters at 463 West Street. I had always imagined that such a renowned mathematician would reside in a palatial office with lesser people scurrying around him. But no, Steve was actually *sharing* an office with several other people who treated him like an ordinary human being. And then, a little later, we all went out together, on *foot*, for lunch in a little restaurant in Greenwich Village. I thought someone as famous as Rice would be carried by a company limousine. Well this was just one of my European prejudices that took some time to subside.

Research Accomplishments

As auspicious as my beginning at Bell may sound, it was, in retrospect, rather pedestrian. My first speech research job at the Labs, suggested by my supervisor Ralph Miller, was *inverse filtering*: you spoke a vowel sound into a microphone, displayed it on an oscilloscope and adjusted tuneable filters until the formant "wiggles" disappeared and a relatively smooth waveform remained, which was a pretty good approximation of the glottal waveform. Neat! As I learned much later, in acquiring more electrical engineering expertise, I was putting "zeroes" on top of the "poles" in the complex frequency plane.

Of course, if the speech signal is the least bit noisy, inverse filtering will only lower the signal-to-noise ratio (SNR). In order to *increase* the SNR, you have to use *matched* filters, just the opposite of inverse filters, piling poles on poles. So inverse filtering is no good for pitch tracking for low-quality telephone signals, which was one thing we were trying to do. But it was an interesting introduction to speech signals.

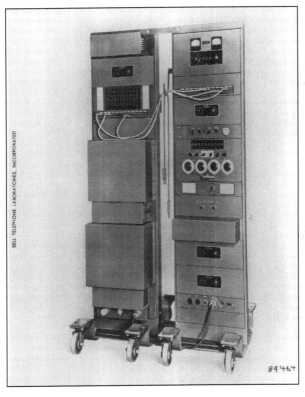

The vocoder. (Reprinted with permission of Alcatel-Lucent USA Inc.)

My next (self-inflicted) project was a 6-channel Vocoder ("voice coder"). Even though a complete novice, I felt that ten or sixteen channels for a Vocoder was more than was really required; I figured that six channels would be enough. Indeed, I thought single-tuned filters would be sufficiently selective if I added some "lateral inhibition" (as in the human eye) to my filter bank: subtracting from each filter a small fraction of the outputs of its two neighbors. It took me many months to build this contraption because I had to wind my own coils. Imagine a Ph.D. in physics, winding coils and spending his days soldering sickly circuits — the technical assistance that I had been promised *presto* did not materialize until a full year later.

Be that as it may, after all the manual labor and mental gymnastics I had invested in my Vocoder project, I was convinced that it produced not only *intelligible* speech but high-quality, human-sounding speech. Thus, I alerted my supervisor, who in turn alerted our director, Win Kock, who — if you can believe it — alerted the president of Bell Labs, Mervin Kelly. They all filed into my lab and the demonstration began. In retrospect, I am convinced they hardly understood a word of what my Vocoder said. But everybody was very polite and all left with best wishes for the future of my project, an early attempt at speech compression, which later became crucial for the Internet and mobile telephony (cell phones). Once, a cooperative student from MIT, Tom

Crystal, played me *his* speech compressor in a noisy, fan-infested, lab. He thought it sounded perfect and so did I. So I suggested that we listen in a very quiet location and I put earphones on — and it sounded horrible.

Another horrible contraption I built was an analog Fourier-synthesizer capable of generating periodic waveforms with thirty-one harmonic frequencies. The gadget, built by the able Herb Hines, comprised almost 1000 little trimmers (variable resistors) and a number of highly unreliable transistors (this was around 1956!). Nevertheless, I got some nice waveforms that helped me solve the problem of minimizing the peak-factor of periodic waveforms by choosing the proper ("Schroeder") phases and making synthetic speech sound more pleasing.

Manipulating just the phase angles of a periodic signal while keeping its power spectrum fixed, I could elicit some astounding perceptual effects including *flat-spectrum* intelligible speech (with H.W. Strube). Later Schroeder phases became important in hearing research — in studying the motions of the basilar membrane in the inner ear.

I also concocted cross-correlation and autocorrelation Vocoders. The autocorrelation function squares the power spectrum. To undo this spectrum squaring, I employed square-rooters, one for each formant range. Only much later did I learn that the proper way to do this is by taking the *inverse* of the correlation matrix (as is done in linear predictive coding of speech and present today in every cell phone).

Although I had been raised in Germany under Hamming's motto "Don't compute—think!" I was soon one of the heaviest users of digital simulation including, with Ben Logan, a "Harmonic Compressor" that comprised 400 narrow bandpass-filters and could be used to slow down natural speech. We made this available to the American Foundation for the Blind, which used it in their "recorded books" program. Of course, the simulation didn't run in real time (this was in 1961) and some eyebrows were raised over my simulations that took several hundred times real time. Even earlier, in the late 1950s, I had started simulating frequency responses of concert halls: a complex Gaussian process in the frequency domain. To accumulate enough statistical data on certain parameters that couldn't be calculated analytically, I let the computer run entire weekends. At $600 an hour, the cost was of course astronomical. But the only financial consequence to me was that my batch-processing budget was raised. Talk about "the good old days" at Bell Labs! I never had to write a single research proposal in all my years at Bell — just do it and, if it made sense, the money would be provided, unless it was "additions to the plant", which were strictly budgeted. What helped of course was that, before "divestiture" (1984), AT&T — within reason — could write off Bell Labs as a business expense.

After the frequency-domain simulations, I began concert-hall simulations in the time domain, in other words, convolving the music signal with the hall's impulse response. The final output was, of course, reverberated music. When we first did this we had a little amplifier with a loudspeaker standing behind the digital-to-analog converter to monitor the computer output. When the people who ran the converters heard the music coming from their machines, they thought some prankster was trying to fool them. But, no, the music *did* come out of the computer — something entirely new in the 1950s.

Later I began simulating 16-channel Vocoders, again in very "unreal" running times. Some people thought I was crazy. They had envisioned simulations for simple waveform coders, such as delta modulation or differential pulse-code modulation. But that was Bell Labs research: you could do "crazy" things if you so felt (and didn't mind a few raised eyebrows). A good case in point is Peter Denes's successful advocacy of *dedicated laboratory computers* for speech research — in defiance of considerable doubts by higher management who at the time (ca. 1964) were still wedded to "computing centers" with big machines. In fact, we sometimes had to travel to IBM headquarters on Madison Avenue in New York City with a trunk full of punched cards representing a single sentence of speech (three seconds).

By 1966, I had become totally disenchanted by Vocoder speech quality. I thought that instead of the "rigid" encoding of the speech signal practiced by Vocoders, we should look for methods of speech coding that left "room for error." At this point I remembered the work of my friends in the picture-coding community, especially an engineer named Ernie Kretschmar, a refugee from Nazi-Germany — in fact the Münsterland, my native neck of the woods. The method of choice in video coding was, of course, prediction: point-by-point, line-by-line, and frame-by-frame. I thought something like that should also work for speech signals, except that the predictor parameters should change with the speech sound; adapting to the different spectra of different speech sounds. That is why Bishnu Atal and I called our first paper *Adaptive Predictive Coding*, later called *linear predictive coding* (LPC) which allowed the excitation signal to be transmitted by roughly 2 kb/sec, an astonishing compression factor made possible by perceptual coding that exploited the masking properties of the human ear.

I was attracted by the opportunity (pioneered earlier at Bell Labs by Ken Knowlton, Frank Sinden, Ed Zajac, Bela Julesz, and Mike Noll) to use the computer-cum-CRT plotter to generate graphic output that no human could ever hope to draw by hand. Number theory was a great inspiration for my computer graphic endeavors. Shows at which my computer creations were exhibited included "On the Eve of Tomorrow," held in Hannover in late 1969 (organized by Käthe Schröder), "Some New Beginnings" in Brooklyn (also in 1969), and "Cybernetic Serendipity: The Computer and the Arts" (by Jashia Reichardt, London 1968).

With Gerhard Sessler and Mohan Sondhi, I constructed an "Acoustic Camera," which was capable of reconstructing a physical object from its blurred acoustic shadow. With Mohan, I also did a study of the restoration of sharp images from pictures degraded by linear-motion blur, and of computer-ray tracing in the deep ocean. This was in connection with what I called "Volume Focussing" (suggested to me by the self-steering antenna arrays that keep communication satellites focussed on the proper earth station). Volume Focussing is capable of focussing on a specified volume in the ocean, important for the defense against submarines. The surface sound channel, called SOFAR, allows sound waves in the ocean to travel over thousands of miles with little attenuation. In a peacetime application, the SOFAR channel is broadcast from Heard Island in the Indian Ocean to Canada, Australia, Antarctica and California to monitor ocean temperatures (and the effect of greenhouse gases) over vast distances.

Philharmonic Hall, New York

In September 1962, Philharmonic Hall at New York's
Lincoln Center for the Performing Arts had opened
to great fanfare with a gala concert under the baton
of Leonard Bernstein in the presence of the First Lady
of the Realm: Jacqueline Kennedy. But audiences, not
least the *New York Times* music critic Harold Schoenberg, were less than enthusiastic about the new hall's
acoustics. For example, Schoenberg complained about
a persistent echo. I got his seat number and when we
looked into this, we discovered that there was indeed
an echo. It was easily taken care of. Ironically, most
seats were free from echoes.

In its predicament, Lincoln Center turned to
a trusted friend on *Lower* Broadway: AT&T, which
turned to Bell Labs where, being in charge of acoustics
research at the time, it ended up in my lap. A Committee of Four was formed under the chairmanship
of Vern Knudsen, Chancellor of UCLA and one of
the founders of the Acoustical Society of America.
Under a consent decree between the U.S. government
and AT&T (1956), Bell Labs was actually prohibited
from acoustic consulting. So it was decided that I
would limit myself to making the necessary measurements to analyze the hall. And analyze we did, using
computer-generated Hamming-window tone-bursts
as an excitation signal and matched filtering on the
acoustic output from the hall. I was assisted in this

work by Gerhard Sessler, Jim West, Bishnu Atal, and Mike Noll, and by Carol Maclennan, who did the complex computer programming.

Before starting the measurements, I asked the ushers, students of the Julliard School of Music, whether there was one good seat in the hall and they pointed to Seat A15 on the second balcony as really good. So I decided to include A15 in our measurements. The results showed that in the center of the orchestra floor, there was a loss of almost 30 dB between middle notes (750 Hz) and low notes (125 Hz). By contrast, the loss was only 4 dB at Seat A15! I later demonstrated this effect to Cleveland maestro George Szell, who agreed that A15 was indeed superior. Then, after the intermission, the legitimate ticket holder for A15 appeared and Szell and I had to sit on the floor, where it sounded even better to me. But I didn't say anything. Then Szell turned around and said: "Down on the floor here it *really* sounds good!" What is the *cause* of this lack of low frequencies (which made the celli in *tutti* passages nearly inaudible)? By time-gating the responses on the computer, we were able to isolate different reflections and the culprit was identified as the overhead reflecting panels.

In addition to the poor bass response the main problem with Philharmonic Hall (now Avery Fisher Hall) was a feeling of "detachment" from the music. To get at this fundamental difficulty, I was able to persuade the German Science Foundation to underwrite a large-scale study of concert hall quality. My collaborators (Dieter Gottlob and Karl Friedrich Siebrasse) made recordings with a specially designed "dummy-head" in twenty-two major halls, which were subsequently reproduced in a large anechoic space by means of an electronic filtering method. This was an early example of the now ubiquitous "virtual acoustic images."

The main finding of this study, involving thousands of paired comparison tests, was that, for good acoustics, there should be strong early *lateral* reflections — a difficult goal, given that most modern halls are wide and have a low ceiling, which favors sound arriving in the median plane of a forward-facing listener's head. To overcome this problem, I proposed diffusing surfaces based on number-theoretic principles to be incorporated in a hall's design. The most widely used reflection-phase gratings are based on *quadratic residues*. I got the idea during a talk by André Weil (brother of Simone Weil) on *Gauss sums and quadratic residues* which he gave in Göttingen in 1977 in celebration of the 200th anniversary of the birth of Gauss, the "Prince of Mathematicians."

The accurate measurement of reverberation time was also advanced by our work on Philharmonic Hall by using reverse (backward in time or "Schroeder") integration. Ideally, concert hall measurements should be made with music as an excitation signal. This can be done by measuring the modulation transfer functions, both on the stage and in the audience area and forming their ratio. With Atal, I also studied sound decays by means of computer-ray simulation. The results revealed gross inaccuracies of existing reverberation time formulas and the strong dependence of reverberation time on absorber *location*. All of these results benefitted the acoustical industry at large.

Colleagues at Bell Labs

No memories of the good days at Bell would be complete without remembering some of the great people I was privileged to work with like the late John Kelly, a true genius, and Jim Flanagan, who joined my department around 1958. One of the greatest leaders at Bell was the unforgettable John Pierce of satellite communication fame and the father of *Telstar*, the first transatlantic space TV, on July 4, 1963 (see Chapter 1). In fact, Pierce was the most inspiring boss and mentor I had between 1955 and 1971, when he left Bell Laboratories to join his old alma mater, the California Institute of Technology. His writing inspired my own writing — although when I once told him I might call my next book "Number Theory for Almost Everyone," he suggested instead, in his usual acerbic mode, "Number Theory for Almost No One." But while often blunt, John could be quite sweet, too, as when, in 1957, he called my Voice-Excited Vocoder "the first speaking machine that sounds human."

John always impressed me by his honesty. Every time I had something unpleasant to report, he would immediately call *his* boss, Vice-president Bill Baker (see Chapter 12), to relay the bad tidings. I was impressed because the culture in which I had been raised in Germany, honesty — volunteering the truth —could be deadly. Lack of forthrightness, I believe, was in fact one of the root causes for the failures of Kaiser Wilhelm's Germany and Hitler's regime — and could yet undo other governments.

One person from whom I learned a lot about human relations and research administration was my boss' boss, our research vice-president, Bill Baker. Bill would always think long and hard before making a decision — which, however, he often communicated in — to many — incomprehensible language. In other words, he made *you* think hard.

Professor of Physics

Around 1966 I received some "feelers" from a physicist friend (Wolfgang Eisenmenger) whom I had invited for a summer at Bell Labs. Was I interested in a full professorship at the University of Göttingen? I was very happy at Bell Labs but didn't know how to say "no". So I said, "Yes, I am interested." (I mean, how can you possibly say "no" — it would be impolite.)

When the offer was actually made, in 1968, I was glad to accept. I had spent fourteen years at the

Labs in New Jersey and my only way "up" was, well, up: less research and more management, which I detested. Thus, in March 1969, I made a side trip to Göttingen (during a skiing vacation in Kitzbühel, Austria, where I broke a leg) to discuss the situation. Of course, Bell Labs didn't want me to leave, so a part-time arrangement was reached with the state government in Hanover that allowed me to do research at Bell when I wasn't teaching in Göttingen. Thus my transatlantic commuter life began — still ongoing thirty-five years later! My agreement with Bell Labs held until my eventual retirement on September 30, 1987, exactly thirty-three years after I had first set foot on Murray Hill soil.

My teaching duties at the *Institut für Schwingungsphysik* (literally "swinging physics" but actually sound, speech, hearing, and concert hall acoustics) consisted of a weekly two-hour lecture with demonstrations on a great variety of subjects: acoustics, electronics, coherent optics, electromagnetics, computer graphics (one of my hobbies) and applications of number theory (my true love). I really liked preparing the numerous demonstrations. The only topic I didn't care for was solid-state physics. Best of all, I never had to teach or test undergraduates. Our institute was what would be called a graduate school in the U.S. The students were highly motivated, intelligent and often smarter than the professor.

Reflection

My life has comprised two careers, one at Bell Labs and another as a professor of physics. Which did I like better? Difficult to say; I was exceedingly happy in both jobs. While at Bell I basically worked as an applied scientist. As a professor in Göttingen, I became a full-fledged physicist, covering a wide selection of topics. I also became a member of the Göttingen Academy of Sciences, which meets every fortnight in plenary session, with talks on many different subjects in the humanities as well as the natural sciences and mathematics. This has led to a considerable broadening of my intellectual horizon. Still, nothing in my memory equals these wonderful and unique years.

WALTER L. BROWN

Walter L. Brown. (Reprinted with permission of Alcatel-Lucent USA Inc.)

Physics in Virginia

I grew up in Charlottesville, Virginia, where my father was a professor of physics at the University of Virginia. As a result I had very early opportunities to appreciate the beauty of physics and its important connections with engineering disciplines. My father had a major responsibility for teaching elementary physics to undergraduates in both science departments and the engineering school. Starting when I was about ten or eleven, he would take me with him to Rouss Physics Laboratory when he set up the demonstrations for his big lecture classes. I remember, as clearly as if it were yesterday, the experiment with a stuffed monkey that was held by an electromagnet to the top, back corner of the big lecture room. An adjustable pressure air rifle mounted on the lecture desk was pointed directly at the monkey. Just as the ball left the rifle barrel, the electromagnet was released and the monkey dropped. No matter what the pressure was in the air rifle the ball always hit the monkey – sometimes just before it hit the floor. This introduction to concepts of the acceleration due to gravity and of dividing velocity into two components made a lasting impression. Then there were experiments with lenses and prisms made by compressing ground-up ice into molds and experiments with melting using constant heat input to a sample of ice. Those were memorable days with my dad.

When I was almost seventeen years old, my connection with the Physics Department at UVA grew into a summer job. Research with ultra-centrifuges, under the direction of Professor Jesse Beams, was a big focus in the department. I was hired as a lab assistant to a graduate student who at the time was working with a continuous flow ultra-centrifuge of the kind that was ultimately used for separating U235 from U238 for use in the first atomic bomb. The process involved adding a fluid into the top of a long, cylindrical centrifuge and then extracting two components of different mass at the bottom. These were the days before "O" rings – 1941 – and getting reliable rotating vacuum seals was tough. I remember taking the whole assembly apart every morning and cleaning it up very carefully with cotton waste (not lintless!) and then reassembling it with clean grease and fresh seals cut from a flat rubber sheet. I guess that was my first exposure to the fact that research involves a lot of sweat and disappointment as well as a lot of joy when the apparatus works right and produces good looking data – even if the results aren't what was predicated and one has to remain open to the possibility that something even more interesting than what was expected is, in fact, going on.

Lane High School in Charlottesville was a sparkling new building when I was in the 9th grade. There were only eleven grades in the school system at the time. In the year that I was a senior, the system changed to twelve years and those in my class were able to choose whether to graduate in 1941 or to continue for an additional year and graduate in 1942. Being a senior two years in a row was a blast! I had a chance to take some extra science classes in a very relaxed atmosphere. I did a project on photography and presented a paper (with glass lantern slides, of course) in Richmond at a Junior Academy of Science meeting. I had changed girl friends during my first senior year, so the second one had other benefits too.

By the time I graduated from high school, the Japanese attack on Pearl Harbor on December 7, 1941 had caught the US napping and, by June 1942, the US was scrambling to assert itself in its war in the Pacific. I had applied and been accepted for admission to Davidson College, a good liberal arts college with a Presbyterian heritage. I had never visited there, and I didn't apply anywhere else. I didn't want to go to UVA. I didn't want to live at home and, besides, my father might give me a really hard time in physics, which I had already decided to make my career.

A Navy Influence

Davidson was a good choice, as it turned out. I only spent one year there and midyear – I was now eighteen – I had a chance to join the Navy V-12 program, which was designed to have its members finish their undergraduate degrees and then become ensigns. In June 1943 the Navy sent me and a bunch of other guys in white Seaman Apprentice uniforms to Duke University. I was now a physics major. All V-12s went to school three semesters per year to speed up the process. The chairman of the Physics Department at Duke was Professor Walter Nielsen, a wise, kindly and perceptive physicist who took several of us young navy guys under his wing and saw that we took the right courses and the right labs to give us a good start. Because it was wartime, the number of faculty members was depleted, but that was not a problem for us. I had a course in electronics (very practical, hands-on stuff – all vacuum tubes and mercury rectifiers and mechanical relays) taught by a postdoc who had just gotten his PhD. I even took a course in physical chemistry and it was a real learning experience.

One day someone tipped over a gas cylinder in the hall and it rocketed down the hallway and out through the end of the building. In one lab, when we were studying the effect of the composition of organic mixtures on their boiling point, my lab partner and I had finished the experiment and poured the benzene-napthalene mixture into the big sink in the lab bench. Another lab group had just lighted their Bunsen burner and tossed the match into the sink. The result was spectacular. There was a big flash and a very short fire and the whole lab was filled with a fine black dust of combustion that gradually settled on every surface. Physical chemistry was fun, but I didn't like it nearly as much as physics. The motion of coupled pendulums, the velocity of a water stream coming from a hole near the bottom of a full bucket, the ionization of mercury by electron impact in a mercury thyratron: that was great stuff!

Because V-12 students went to school at Duke three semesters per year, our class graduated in June of 1945 and headed off for Midshipman School. The plan from the beginning was that we'd take three months of *real* Navy training: ships, guns (from 50- caliber machine guns to the 16-inch guns on battleships), aircraft, torpedoes, gun directors (amazing electrome-chanical devices), communication devices (from flags and flashing mirrors to radios and radar) – the works. Then we'd be tested and, if we passed, we'd become ensigns and exchange our white middy blouses for officers' uniforms and begin to do something useful for the Navy. But the war was over before we gradu-ated from Duke and it was clear that we would all be discharged from the Navy sometime during the next year. In that last semester at Duke, Dr. Nielsen worked hard to help those of us who wanted to con-tinue our education to apply to graduate schools with the expectation we probably would not be entering until the fall of 1946.

The Navy sent me to Midshipman school at Cor-nell (high above Cayuga's waters). I suffered an injury to my knee when I was on a short leave in Charlot-tesville between Duke and Cornell, it became infected after I got to Cornell, and I spent a couple of weeks in the infirmary and missed my class, getting started with the next class about a month later. We did learn a lot about the *real* Navy, including seamanship on a YP (Yard Patrol) boat on Lake Cayuga from a *real*

Navy Chief Petty Officer. We also learned how slippery the hills at Cornell were when they were ice- or snow-covered. Graduation was on December 7, 1945 and, sure enough, we became ensigns and gentlemen (by decree).

My service to the Navy as an ensign was pretty chopped up: I was first sent to Boston, waiting the start of steam engineering school in Newport, RI. I think the high point in my naval career was the one weekend I spent as the Captain of Old Ironsides in Boston Harbor. I was clearly just decoration. A Chief Petty Officer did everything that had to be done, but I slept in the captain's quarters in the stern of that grand old ship and tried to look presentable for tourists who visited her. While in Boston I observed a very practical example of the physical quantity "momentum." A Navy cruiser (not as big as a battleship, but a very big ship) came in to tie up at a huge pier in Boston Harbor (not where Old Ironsides was). The pilot – maybe it was the captain, but not likely – wasn't paying close enough attention and the ship didn't slow down soon enough. Although it was moving very slowly, when it reached the dock it plowed ten feet into the pier before it stopped. Very impressive! An even more valuable experience while I was in Boston was a class in machine-shop practice that I was assigned to take. I had done woodworking at home in Charlottesville but had never machined metal. So I learned about grinding tools to cut properly in a metal lathe, and how to use a milling machine, and how to clamp work before drilling holes so you wouldn't damage the piece or yourself, and the difference in the edge of a drill bit for drilling brass or steel. Very helpful stuff for an aspiring experimental physicist.

Then it was off to Newport, Rhode Island. The second day of steam engineering class I received new orders: I was to report to the War Plans Officer at the Bureau of Navy Personnel in Arlington, VA. I had the demanding job of taking a different volume of war plans to a Captain (the WPO) each morning and returning it in the afternoon. My other responsibility was to keep track of the personnel who entered the different Boot Camps in the US, how they progressed in their training, and when they left. I made one interesting discovery: more guys were going into the Great Lakes Training Center in Chicago than were coming out. It turned out the Commandant of the Camp was keeping them to take care of the golf course. It made a minor stink.

I was next assigned to the Office of Naval Research in downtown Washington. Real science! This organization was one of the most farsighted things the government did. The office was responsible for assessing research proposals from US universities and making grants for their research. That funding source came very early in the business of government support for university research. I was, of course, a babe in the woods as far as assessing the new ideas of some of the best scientists in the country, but I had a chance to learn about some of the exciting things that were going on, or would be going on, at the forefront of physics. I had that assignment until August of 1946, when I was discharged from the Navy.

Graduate School

On to graduate school. Professor Nielsen was a friend of the chairman of the Physics Department at Harvard, Professor James Van Vleck (later to win the Nobel Prize). It was Nielsen's recommendation of me and another V-12 Physics major that resulted in our acceptance to graduate school there. Many of the starting graduate students in the fall of 1946 had been in military service and we all felt the challenge of graduate courses after a hiatus from formal university education. Some, with wives and several children, were very highly motivated to move as quickly as possible through to a Ph.D. I had married Lucie Oakes in June 1946, two weeks after she graduated from Duke, and we had a marvelous four years in the Boston area, working hard but enjoying a whole world of new experiences: living as governess/cook/maid and handyman/butler with a wealthy family in Newton Center and, in the summer at their big home on Buzzard's Bay; then, when I started my Ph.D. research, in an apartment on the first floor of a three family house in north Cambridge with a single bicycle for dual transport to supplement the MTA. We made many extraordinary new friends and Lucie had many new work experiences, including being companion to the mother of the president of Harvard.

I was fortunate in being accepted as a Ph.D. student by Professor Ed Purcell, a Nobel Laureate in 1952 for his work on nuclear magnetic resonance and a physicist with an incredible gift as a teacher. He had the ability to recognize where a student was in his or her understanding of a problem and what the conceptual roadblock was that was limiting further progress. He could shift discussion of a problem to another plane with an analogue that would turn one's inner light bulb and cause one to say, "Oh, now I see." Purcell wanted me to utilize nuclear magnetic resonance to make a measurement that would calibrate the fundamental dissociation energy of the deuteron. I made a precision determination of the energy of a gamma ray from Thorium 242, by measuring its internal conversion electron in a 180-degree magnetic spectrometer that I machined and whose magnetic field I measured with nuclear magnetic resonance. In addition to Purcell, I had important help from Professor Bob Pound, an electronic whiz who had assisted Purcell on his Nobel-Prize-winning work, and Professor Ken Bainbridge who was an expect nuclear physicist and had worked on the Manhattan Project. The period 1948-1950, when I was working in that lab, were wonderful years. I successfully defended my thesis in the fall of 1950.

Bell Labs Calls

In the early months of 1950, I began interviewing for positions, anticipating that I would finish my PhD early in the summer. Among interviews with a variety of universities was one with Dr. Deming Lewis from Bell Telephone Laboratories. (After he retired from Bell Labs he served as president of Lehigh University for several years.) That interview resulted in my receiving an invitation to visit Murray Hill for further interviews, and on an icy day in February, I drove/slid in a 1948 two-tone Chevrolet sedan down the Wilbur Cross Parkway from Massachusetts through Connecticut and New York and on to Summit, NJ. I don't remember much about my talk about my thesis work, which was clearly not in a field of interest to Bell Labs. In those days, Bell Labs was hiring people that were judged to be pretty smart, well-educated and flexible in the field in which they would work. I talked to a number of Bell Labs scientists and engineers and a few were particularly memorable.

J. R. Pierce, otherwise known as J. J. Coupling – his pseudonym in his science-fiction writing – was unknown to me before I met him in his office. A highly impressive man from the very first meeting, with penetrating questions about why I had done what I did for my thesis and what did I think I wanted to work on next (hopefully not a continuation of that work). He was also friendly and clearly smart as blazes. I guess the thing that characterized Pierce, as I knew him better in later years, and that certainly left an indelible impression on me that day was his file cabinet. It was a standard-size, five- drawer file cabinet. The drawers were labeled: TOP DRAWER, NEXT TO TOP DRAWER, MIDDLE DRAWER, NEXT TO BOTTOM DRAWER, BOTTOM DRAWER. A very clear filing system for a man who invented the Pierce Gun for vacuum tubes and the reflex klystron as a high frequency oscillator and who was to become a pioneer in space communication systems.

Lester Germer was another very special person whom I talked with. Dr. Germer had worked with C. J. Davisson in the late 1920s, when they discovered electron diffraction while bombarding a nickel crystal with an electron beam. For that work, which was always called the Davisson-Germer experiment, Davisson shared the Nobel Prize in Physics in 1937 with G. P. Thompson of Cambridge, England, who had discovered electron diffraction in a different experiment at the same time. I always wondered if Germer had felt slighted in that award. In 1950 Germer was doing research on the problem of electrical contact failures in mechanical relays, the heart of the Bell System switching network, a tough problem. He was a very intense, thin man who told me that his doctor wanted him to gain weight by putting him on a diet heavy in peanut butter, but it didn't work.

It was clear that Bell Labs was full of bright, highly motivated, hard-working and creative people, and it seemed as though they had a lot of opportunity to do research without their bosses hovering over them all the time. When I told my dad about my visit, he confirmed that Bell Labs had a terrific research reputation and hoped I'd get an offer to go there.

In a couple of weeks, I had a call from Deming Lewis extending me an offer for $5000 a year, and asking me to report for work by the end of the summer. My wife and I discussed the pros and cons of Bell Labs compared with university offers, and I accepted and dove back into Lyman Physics Lab to complete my thesis work. There were snags in wrapping it up and I wasn't finished by the end of the summer. Fortunately, Bell Labs was generous in extending my report date, and I finally showed up at Murray Hill on December 1, 1950. With the help of the Bell Labs housing office we rented an apartment at Gales Drive in New Providence. There was quite a bit of insider dealing between the head of housing and apartment managers, and we always wondered about how waiting lists worked.

A number of other new employees arrived about the time that I did, and we were given a week-long orientation course on the Bell System and Bell

Telephone Laboratories at 463 West Street in New York, where most of Bell Labs had been before the new Murray Hill lab was built. Ride on the Lackawanna RR through Summit to Hoboken, ferry across the harbor, walk to 463 West Street. It was usually a stimulating commute for a new recruit except one day when it was raining. I recall being impressed by the fact that a freight railroad ran through the back of the building on elevated tracks. I don't remember much that we were taught, but I do recall a message from W. S. Gorton who lectured us on the importance of keeping a good, registered laboratory notebook and writing down – in chronological order, in ink, with no erasures, with cross-outs if necessary – everything we did or thought about so we'd have a reliable basis for patent applications; and then having it witnessed by your boss. He said "The palest of ink is more reliable than the most retentive memory." It turned out that Gorton, early in his career, was the inventor of a particularly effective vibration suppressing galvanometer support, and one of the labs that I was later in, the lab where Walter Brattain did his 1947 point contact transistor experiment, had one hanging from the ceiling. Attached to it was a translucent circular arc, about a meter long, with a precision millimeter scale etched into it. Galvanometer? In that lab, and in many others, there was a moveable stand, about 3-feet high from the floor, with a black wooden top about 18-inches square. This was to put your notebook on, so you could record data on the spot – no loose pieces of paper.

With the indoctrination over, my first assignment was to work with Germer on relay contacts. What materials would provide the most reliable contacts, taking into account the arcing that took place as the contacts closed and opened? But this was in the winter of 1950-51, and the Korean War was heating up. When I was discharged from the Navy in 1946 it was into the naval reserve and I was eligible to be called up for active duty. To avoid this possibility, early in 1951 Bell Labs transferred me from Germer's department to Bill Shockley's department. Work on the newly invented transistor was such a hot topic that the government declared it essential and people working on it were exempt from being called up for military service. So I was suddenly a novice among giants in this new field: Dick Hayes, Gerald Pearson, Willy Van Roosbroeck, Harold Montgomery, Bert Moore in my immediate group; Bruce Hannay, Ted Geballe, Frank Morin, Howard Reiss, Jim Struthers, Bill Pfann, Bob Batterman and many more in the semiconductor chemistry and materials group. Of course Bardeen and Brattain were in other departments nearby. Relations between Shockley and Bardeen and Brattain were pretty strained since 1948, as I learned pretty quickly, and it wasn't reasonable for B and B to have Shockley for their boss as he had been earlier. Shockley was incredibly smart and inventive and he had an ego that was incredible too. It was awesome to work for him – I never really felt as though I worked with him. He was so full of ideas and had such deep understanding and theoretical and experimental insight that it was a challenge to discuss new experimental observations and their implications with him. He suggested I look into the electrical properties of germanium single crystals near their surface. Germanium was the semiconductor material of choice at the time. Because it had a much lower melting temperature than silicon, its purification and high-quality, single-crystal growth had been accomplished long before silicon was a viable alternative. Germanium crystals doped p- or n-type were available and the techniques for cutting, polishing and making contacts to samples of a desired shape and size were too.

Measurements of the surface conductance and surface capacitance of germanium, as altered by the voltage on an electrode placed close to the surface, led me to the first observations of the field effect on surface conductance and the inversion layers at the surface of a semiconductor, and the sensitivity of both of these properties to the chemical environment of the surface. Wow! Exciting stuff!! The results were published in *Physical Review* in 1953, probably the most important publication of my career. Those effects in silicon are the bread and butter of modern silicon devices.

I then turned to the study of the effect of bombardment of germanium by alpha particles. Joe Burton and Ken McKay, in different departments of research, had previously worked on alpha-particle bombardment of germanium and diamond and observed electrical impulses from the impact of single particles. One of the great features of Bell Labs was that there were so many experts in so many different fields within it. Furthermore, they were all accessible and glad to share their experience and knowledge with others. The openness of the intellectual atmosphere was a huge strength. The management of the Labs had certainly gotten that right. Work inside Bell Labs and by others outside of it, led to the development of remarkable particle detectors made of germanium and of silicon, an enterprise important in my own department at Bell Labs in the 1960s. More about that later. My goal in the 50's was to observe how the properties of germanium were altered by repeated bombardment with many particles. To simplify the physics, I soon turned to bombardment with high-energy electrons rather than alpha particles. Above threshold energy of about 300 keV, an electron has enough energy in a collision with a single germanium atom to displace it from its position in the crystal lattice and move it to an interstitial site, thus producing two defects in the crystal. Defects influence the electrical properties of the material because either electrons or holes can occupy them. Walt Augustyniak, a terrific assistant,

and I explored the formation and annealing properties of defects created by high energy electrons, first
using electron accelerator facilities at MIT and then
at Murray Hill after acquisition and installation of
a 1 MeV electron accelerator in Building 2. By this
time Shockley had left Bell Labs and I was part of
the Semiconductor Research Laboratory under A.
D. "Ad" White and in the Semiconductor Physics
Research Department under Joe Burton. Joe was
my supervisor and mentor for many years. He was
an enthusiastic proponent of pursuing new ideas in
the open research environment of Bell Labs, and so
was Ad White, and so was Bill Baker who was Ad's
boss as Vice President, Research. The "dive into the
woods" admonition inscribed in the reception hall of
Arnold Auditorium was real.

Research in Nuclear Physics

Late in the 1950s Joe Burton had a new idea that he
tried out on me. Nuclear physics was a field that Bell
Labs was not contributing to, but it was an extremely
active area of research in the US and throughout the
world. Wouldn't it be valuable for us to add that to the
breadth of research at Bell Labs? Although my PhD
was in a little corner of nuclear physics, I didn't have
much intellectual depth in the field and yet Joe's idea
sounded good to me and Ad White and Bill Baker
agreed. Soon after that, in 1958, we hired Paul Dono-

van, a genuine nuclear guy from Berkeley, and early in 1959 Walter Gibson, also from Berkeley, joined the new effort. By this time Bill Baker had reorganized the research area. Ad White had been promoted to Director of Research – Physical Sciences, Joe Burton had been promoted to Director of Semiconductor Research, and I had been promoted to Department Head of Semiconductor Physics Research. It was in this department that the nuclear physics effort grew and flourished over the next twenty years, with heavy emphasis on connections with solid-state science and with computer science.

Donovan and Gibson were pioneering gurus in the development and utilization of extraordinarily effective silicon semiconductor particle detectors, using the strength of Bell Labs in work with semiconductor materials. They used them in their own nuclear physics research and shared their expertise with commercial efforts that grew up rapidly. Nuclear physicists were very early in recognition of the enormous importance of computers in experimental measurements and our group made valuable contributions to the development and application of what now seem like very primitive machines designed to handle the multidimensional data of leading edge experiments. In the early days the experiments were done with nuclear physics facilities at Brookhaven National Laboratories.

After a few years we had established a collaborative program with Rutgers University that enabled the University to acquire a state-of-the art accelerator while providing us with first class facilities close to home. That productive collaboration continued for twenty years, and involved half a dozen scientists from Bell Labs, a comparable number from Rutgers, and a series of graduate students who completed their PhD's with Bell Labs thesis advisors. Amazingly, Bill Baker kept track of all of us — expressing enthusiasm for what we were doing, calling each person by name and identifying his work when he encountered one of us in the hall.

Telstar

In 1960-61 a new enterprise challenged Bell Labs. It was the Space Race, in particular the race to demonstrate the feasibility of a broadband satellite communication system — *Telstar*. At its peak I think a third of all the people in Bell Labs must have been making contributions to this hallmark Bell System program. It was wonderful — surely the most spectacular effort that I was ever involved in. Only Bell Labs could have pulled it off with the speed and technical quality that it had, and it worked; even if the British did have the wrong polarization of their antenna on July 10, 1962 so that the French scooped them in that historic transmission from Andover, ME. Our department was engaged in it with big help from Tom Buck and George Wheatley in Jim Lander's department and from Laurie Miller, a consultant from Brookhaven National Laboratories who was a whiz at nuclear electronics. We designed and built a cluster of detectors and associated electronics to measure the electrons and protons that had been found by Professor Van Allen of Iowa, to be trapped in the magnetic field of the earth – in the Van Allen belts. Telstar was to fly through the regions containing these particles and we wanted to know how serious they were as damaging agents for semiconductor electronics including the solar cells that provided power for the satellite. We had the right mix of experience for the job: knowledge of damage in semiconductors due to energetic particles and of design and construction of particle detectors appropriate for measurements of Van Allen particles. It was a blast!

Telstar *groundstation, Andover, Maine. (Reprinted with permission of Alcatel-Lucent USA Inc.)*

I was at Andover when *Telstar* was turned on for its first transmission and when it performed magnificently the roar of the crowd of Bell System people present was enormous! It turned out that, on the day before *Telstar's* launch, the US had set off a high altitude nuclear explosion that filled the Van Allen belts with a very high density of high energy electrons – much to the surprise of the bomb designers. They were a challenge for us to measure, detrimentally affecting the transistors of the satellite, and pressing the satellite circuit designers to keep *Telstar* operating. But the Bell System had a marvelous winner. All of Bell Labs basked in the success. For me and my wife 1961 and 1962 were a whirlwind – our fourth child was born, and we designed, built and moved into a new house. That activity was intermixed with me running back and forth among Hillside (the satellite construction site), Whippany (the satellite component test site), the Goddard Space Flight Center (NASA's satellite integration site) and Murray Hill (where I was, among other things, the phantom foamer early in the morning for modules of our electronic package). I'm glad I was only thirty-seven or thirty-eight at the time, but overall it was a wonderful and unique experience.

The Space Race expanded at Murray Hill with the hiring of Lou Lanzerotti to help understand the particles-in-space science and its engineering implications for the Bell System. With the worldwide connections he forged, this activity continued with a steady stream of visitors and collaborators, and government support even after my retirement in 2002. The last vestige of the experimental efforts in this program at Murray Hill was still in place in 2008 on the southeast corner of Building 1 – a radiotelescope that tracked the sun and studied its interfering high-frequency microwave emissions, part of Lanzerotti's original program.

Ion Implantation

During the evolution of our nuclear physics program, the interaction of ion beams with solids became a growing interest. Some of the work required the relatively high- energy accelerator available at Rutgers. Beautiful experiments were done studying the channeling of high-energy ions in single crystals along axial and planar "channels" formed by rows and planes of atoms that the ions encountered in favored directions. However, lower energies were of interest too, and we acquired first one, then a second, and finally a total of four accelerators in my department and our sister department headed by John Poate. They were used for channeling measurements of the location of impurity atoms in crystals, of the level of damage in crystals shown by disruption of the orderly array of crystalline atoms, and also for the new field of ion implantation. This latter field involved changing the properties of solids relatively near their surfaces by "implanting" ions into them. These studies, and similar ones carried out by Al MacRae's group in the development department, led to the universal use of ion implantation in silicon IC factories as a standard means of adding specific quantities of specific dopant atoms to silicon (and other semiconductors) to control their electrical properties.

The three accelerators that had MeV capabilities provided another, very different value. When helium ions of about 2 MeV impacted a solid, a small fraction of the ions made nearly head-on collisions with the nuclei of atoms of the solid and bounced back with an energy less than they had to begin with. The reduction in energy revealed the mass of the atom that the ions had collided with, and at what depth. The energy of the ions was measured with silicon semiconductor particle detectors (the kind we had helped develop in the late 1950s and early 1960s). This Rutherford Backscattering Spectroscopy (RBS) was a marvelous method for probing the depth dependence of the composition of a solid within the first few micrometers of the surface. Since many new technical innovations involved thin films of this thickness, the analytical strength of helium ion backscattering was used in countless measurements of thin films of specific interest to us and to others throughout Bell Labs. At one time there were the three accelerators in the research area and two in the development area (one at Murray Hill and one at Allentown) that devoted a large part of their time to this kind of measurement.

One thing RBS lacks is the ability to examine very small regions of a surface and as interest in the properties of structures and devices with sub-micrometer size expanded, there were increasing numbers of cases in which RBS was not applicable. Nevertheless, ion beam-solid interaction activities continued at Murray Hill with slowly diminishing emphasis, with fewer accelerators and with many variations until 2002.

Last Thoughts

In 2002 Agere Systems, formed by spin-off from Lucent in 2000, gave up its research in physical sciences, and the last and newest of the accelerators was sold, and the last of the research workers, including me, were laid off or retired.

It had been a fantastic run of just under forty years since the first ion accelerator was installed at Murray Hill. The people who made the biggest impact in it were Walter Gibson, Len Feldman, and John Poate, all marvelously creative and productive scientists and all extraordinary mentors of people who worked with them. It was my privilege to have had them in my department and to have been able to help facilitate the ideas that they had. There were an enormous number of visitors who came to Bell Labs to work with us in this ion-solid field, some of them graduate students from Cornell, Cal Tech, Harvard, the University of Virginia, and the University of Catania, carrying out PhD work, part or all of it at Murray Hill. There were visiting professors from South Africa, Denmark, Sweden, Australia, England, France, Germany, Netherlands and Italy as well as the US who came for stays of a month to a year. It was an incredible environment that lasted through the AT&T divestiture in 1984 and the formation of Lucent in 1996.

During all of this time Bill Baker maintained a presence at Murray Hill and when I encountered him, only occasionally since the mid 1990s, he was the same as always – remembering me, identifying me with past projects, and inquiring about the present. A remarkable man just as Bell labs was a remarkable place.

CAROL MACLENNAN

Carol Maclennan at McMurdo Station (Antarctic) with Dan Dietrich (L) and Larry Lutz. (Reprinted with permission of Carol Maclennan.)

Note: Carol Maclennan was interviewed by Sheldon Hochheiser on 29 September 2008. The interview was then edited, retaining Maclennan's words.

From Pembroke to Bell Labs

When I was a child, my family lived in Bound Brook, New Jersey. My father, originally a Canadian, was a chemist with a Ph.D. from NYU and worked for what would later become American Cyanamid. He came to the US to go to graduate school, got a job here, and met my mother. I was the eldest child and once I was born she stayed at home with me. I had a sister who was born a year and a half later and another sister born ten years later. All girls.

I always liked math and was a math major at Pembroke College. My mother had gone to Barnard College and I think she would have liked very much if I had gone there too. However going to a university in New York City didn't seem to suit me. Of course Providence probably wasn't that different, though smaller. Pembroke was part of Brown University at that time, and the campuses were slightly separate, down the street from one another. Only a few classes, like gym, were not coed. Pembroke was small compared to Brown at that time – maybe one woman to five guys. The math classes had at least that ratio, although there were a fair number of math majors at Pembroke. In general Pembroke women were better students than the Brown men. Or perhaps that's just my prejudice.

My interest in math was always encouraged. Part of that is probably due to my dad being a scientist. And we certainly were not discouraged at Pembroke and were encouraged to do whatever we wanted to do.

When I was finishing at Pembroke, I considered graduate school, but I felt that I was ready to do something other than school for a while. Towards the end of my junior year, Elaine Lewis came from Bell Labs to recruit women for a summer program. She was a one-woman voice at that time in encouraging women in science at Bell Labs. Way ahead of her time! There were maybe twenty-five of us who ended up in the program that year. When I said at the dinner table at home – while on a vacation I suppose – that I had this interview with someone from Bell Labs and that there was a possibility of a summer job, my father said, "If you can get a summer job at Bell Labs, you should take it." He knew about Bell Labs, being a scientist, and probably knew people there. That was terrific advice, and of course I probably would have taken it anyway, but I was very pleased that was something that he thought was important.

That summer I worked in the acoustics department learning FORTRAN. That was the computer language that everyone used. I worked with Ben Logan. He was an engineer and mathematician with very good intuition about a lot of different things. He was also a blue-grass fiddler and was known as Tex Logan in the country music field. He was an interesting guy. I did some simple programming, got to know the department and had a good time. It was the first time I worked with computers.

When I graduated a year later from Pembroke I decided to go back to Bell Labs, and I went back to the same department, acoustics. I was called a senior technical aide (STA). I worked with mostly the same people with whom I had worked the previous summer. I did some work on the block diagram (BLODI) compiler. Engineers would draw little pictures of block diagrams, and the computer would turn it into machine code, mainly for speech processing. This was pretty advanced for the time. Some of it may have been done in FORTRAN; although I'm sure there were basic machine language subroutines as well. The researchers in the acoustics department liked to use this program, and they would want new little boxes that would do something different. That was what I was doing, trying to expand it for them. They would want to run various speech data through it, so I did that too, sentences about "yellow dogs" and "roaring lions."

Women at Bell Labs

I was not the only woman in my area at Bell Labs at that time; I can think of maybe ten others, but that was probably over a period of time. There were some psychologists and a couple other programmers. I shared an office with a Czech woman at one point. She knew a poem or saying in Czech where the number of letters in each word was the next digit of pi, and she could get pi to about a hundred places. There were not a lot of women that were at the senior level of Member of the Technical Staff (MTS). Even some women with Ph.D.s were not MTS. Of course I was not a Ph.D. I particularly remember one psychologist woman who was a Ph.D. and was only an associate member of staff (AMTS).

I'm not sure I thought in terms of career at that point. I was in my early twenties. The first thing I wrote myself was something about the BLODI compiler. It was not published in Applied Physics Letters or anything. It was an internal memo, a Technical Memorandum (TM).

I was married when I got out of college, and my husband was in graduate school at Rutgers. Then he went to Cornell for a PhD, so I moved up there and got a job at the computing center programming for some professors. It was not the computer science department; it was the department that did computing for people in various places.

About a year and a half later I came back to Bell Labs. At that time Walter Brown interviewed me. He had a position open for a programmer to do analysis on spacecraft data. Someone had already filled my other position from a year and a half before. I went to work for Walter Brown in Semiconductor Physics and worked for him for many years. He was terrific, and I enjoyed work in that area very much. *Telstar* had recently been launched, and Walter had a semiconductor detector on it. *Telstar* was launched just after one of the big nuclear explosions in the atmosphere, and the detectors on spacecraft that were up there were saturated. His detector, having been launched after the explosion, got the only good data about the electrons decaying. That was very exciting. There was another woman in the same office. Her name was Lee Davidson. She later left and married a rabbi. At first I worked with a young man named Charlie Roberts. Then I began to work with Louis J. Lanzerotti and ended up working with Lou for more than thirty-five years.

One advantage to having left was that it was much easier to be hired at a higher level than it was to get promoted to it. When I came back from Cornell I came in as and stayed an associate member of technical staff (AMTS) until about 1980. I cannot remember exactly when it was, but the administration at Bell Labs looked at the category and found that the AMTS category was filled mostly with women, many of whom probably should have been MTS. I was not a Ph.D. but I got experience and got better, and by 1980 or 1981 I was ready to be an MTS. They got rid of the whole category of AMTS. Those who had not been at the Labs for very long had to do some educational studies and then they became MTS, but those who had been there more than ten years or so were automatically made MTS. I was made an MTS.

After that, a lot of very good women came into the labs. Several were on my corridor. Alice White, who is still at Lucent and has done very well there, came in as a new Ph.D. Julie Phillips, who is now in New Mexico at Sandia and doing very well. Cherry Murray was on my corridor for a while. She is now

Dean of Engineering at Harvard. I think having more women made it a more amenable place. Women often tend to think about the people involved as well as about the physics. We started having tea in the afternoon – people would come in and talk physics.

I was very active in the Canoe Club for a long time. The Canoe Club existed for many years. When I first joined in the 1960s we had a member named Bucky Buckland who told stories about when he worked at Bell Labs in NYC before Murray Hill was built, and how the Canoe Club was active even then. There was one annual trip that still runs every year to the Adirondacks for cross-country skiing. That today is run by some remnants of the Canoe Club. Of course anyone can go now, but one has to know someone to find out about it. The Canoe Club used to have trip committee meetings two or three times a year. People would contact their friends and acquaintances and we would have trips almost every weekend – hikes, bicycle trips or canoeing. It was not just canoeing; there was skiing in the winter, for example. The Canoe Club was quite active until the 1990s. One of the last things we did was revise a book called *Exploring the Little Rivers of New Jersey*.

There were also "out-of-hours" classes back in the 1960s and 1970s. After hours starting at six o'clock people would teach a class for an hour or so. Anyone could sign up. I don't think we had to pay any money. The Labs provided the facility, and the teachers were generally Labs members. I took classes in Russian and Japanese. I took a class in home maintenance; how to fix the toilet or whatever. I still have some of the notes and use them occasionally.

While I did other things too in my work, I did a lot of spacecraft data analysis. I did not always work with Walter Brown. Lou and I were in Walter's department for many years. Initially I worked with some other people in those departments, but gradually it became that I worked mostly only with Lou. The department name kept changing but usually said something about Semiconductor Physics. After *Telstar*, I worked on other spacecraft data: ATS, IMP, *Explorers*, *Voyager*, and *Ulysses* among others. Lanzerotti was a co-investigator on *Voyager* with some people at Johns Hopkins Applied Physics Lab and other places, so I was involved in a fair amount of the *Voyager* analysis. I did not work on the initial processing but tried to pull things out of the data. We had a good time.

Another thing that happened is I got remarried and then had a daughter in 1974. I met my husband on a ski trip with the Bell Labs Ski Club. He was a mathematician at Bell Labs. I didn't take off much time from work after I had my baby – maybe a month and then I started back to work part-time. Walter and Lou were very amenable to whatever arrangement I wanted to make. Sometimes I brought my daughter in to the Labs. I would pull out the bottom drawer of my desk, and this little being would sleep down there. After a while I got childcare. She turned out well – she's now a history professor at a small honors college in Florida.

But at that time there were no provisions for childcare at Bell Labs; we were totally on our own. There was a group called the Women's Rights Association that asked the senior management to get childcare facilities – or subsidies or discounts for childcare with some local group. We talked to Arno Penzias, who was then Vice President of Research, and he said, "Oh, can't you just hire some lady from Plainfield to come in and take care of your child?" He didn't understand.

Geophysics at Bell Labs

One notable paper from 1973 was the one on pork bellies that I co-authored with Lou and Tony Tyson, published in *Physical Review Letters*. That paper correlated gravity waves with a commodity index. We had to title it "Correlation of Reported Gravitational Radiation Events with Terrestrial Phenomena." Joe Weber at the University of Maryland had measured something he claimed were gravitational waves. Tony was interested in the gravitational waves, and he didn't believe that Weber was really seeing them, so he came to us since we had access to all kinds of geophysical databases. If they were really gravitational waves they should correlate with something geophysical, so we tried to correlate them. We also correlated them with things that obviously did not have any relevance to the gravitational waves. One of those was the pork belly index on the Chicago Stock Exchange. We got as high a correlation there as we did with anything else. That was the essence of the paper.

That was not my first paper. I probably did some earlier with Charlie Roberts. That would have been in the 1960s. I think he left Bell Labs by the 1970s. There were also some on proton measurements from spacecraft. I don't know if I got my name on anything with Walter Brown and *Telstar*.

My role in these projects was to perform calculations on the computer. I sorted out the data and tried to make it make sense in some coordinate system. Many of the papers were discovery papers – data that no one had ever looked at before and in places where no spacecraft had ever been. What's there? What do you see? How do you interpret it?

With *Voyager*, I worked mostly on Jupiter data. That was very exciting, because it was cutting edge. We looked at protons and ions in the radiation belts of Jupiter and how they co-rotated around with the spin of the planet. The planet spins every ten hours. It's humongous, so things are really whipping around, especially what is far out. There was a lot of physics, not all of which I understood.

We were doing spacecraft data analysis and there were not a lot of people doing this at the Labs. It was a very small group. We had a lot of summer students (one year they even had special t-shirts) and regularly hired a high school student to help us out after school. Some of these students went on to study geophysics, and one, Martha Kane, was the second woman to winter at South Pole. We also often had visitors, as most of our colleagues were outside Bell Labs. We collaborated a lot with the Johns Hopkins Applied Physics Lab. For the most part, we did not do a lot of work with other groups at the Labs. If we had statistics questions I often would talk to people in the Statistics Department. Bill Cleveland and Beat Kleiner (from Switzerland) were interested in large sets of data, and we had that. In fact, there was a computer program called "S" that the statisticians had developed. I used that a lot for a while because it could handle large amounts of data and display it in reasonable ways. We had large amounts of data. We also collaborated a lot with David Thomson, an expert in spectral analysis in the math department. Together, beginning in the 1990s, we wrote some very exciting papers about finding modes of the Sun in geomagnetic data.

In the 1980s, we took a magnetometer to Greenland. I was involved because I had written some of the software to run the data system. The magnetometer was controlled by a low-power microprocessor and everything ran off a battery. The software was written in [DEC] PDP-8 code. This was different from anything else I had done, and more like what

I did with the BLODI compiler. It was machine-language. We took those magnetometers to all kinds of interesting places, often in remote settings. I think the first one was somewhere in upstate Vermont. We had a whole chain of these and other magnetometers in people's basements with regular power. We were looking at the plasmapause, an area in the magnetosphere where the plasma density drops off rapidly. I always enjoyed doing fieldwork. Les Medford, our engineer, was responsible for most of that. For various reasons he did not want to go to the Antarctic, and when we needed to have an instrument moved down there I raised my hand and said, "I could do that." He wrote detailed recipes for how to install it and what the meter has to say and what to do if it does this, etc. I went down there and did it. I first went to the Ice in 1984, and went again a total of three times. That was quite an experience!

We had some connections with the *Telstar* people when AT&T reentered the satellite business in the 1980s. We got data from them regularly and had summer students working on some spacecraft charging problems with them. A spacecraft sits in the magnetosphere and gets charged on one side relative to the other. Then it can discharge and nasty things can happen. Data can be lost and the spacecraft electronics can be damaged. Unfortunately the satellite area was divested in 1997 so this subject was no longer of interest. In collaboration with New Jersey Institute of Technology we put a solar telescope on the roof at Murray Hill around 2001. Before that we were trying to correlate cell phone outages with solar radio noise. We had had difficulty getting good solar radio noise measurements so we thought to acquire our own. We did a lot of work with cables too, using the old submarine cables in which AT&T had no more interest. We measured the induced voltages on them and tried to correlate that with various solar effects. We did measurements across the Atlantic, measurements to Hawaii, and worked with the Japanese on some of that. A lot of it was correlated with solar activity. Some of it was correlated with some kind of gravity waves in the atmosphere.

In 1990 a spacecraft called *Ulysses* was launched, and that essentially is what I worked on during most of the 1990s and until recently. The spacecraft initially went to Jupiter, and used Jupiter to put it in a polar orbit around the Sun. Everything else goes around the Sun in the ecliptic plane, but *Ulysses* goes around in a vertical plane, roughly perpendicular to the ecliptic, maybe 80 or 85 degrees. It used Jupiter to swing it out of the ecliptic plane and put it into this orbit. It was originally to be launched much earlier, but when they canceled the shuttle because of the Challenger disaster it got delayed and was eventually launched in 1990. Lanzerotti was the Principal Investigator on this experiment, which was a particle instrument measuring electrons, protons and ions up to iron. It could differentiate from hydrogen all the way up to iron. It could differentiate hydrogen, helium, carbon, nitrogen, oxygen, neon, magnesium, silicon, sulfur, and iron. I worked a lot on that. *Ulysses* is now turned off, but we were getting good data as late as 2007. I was not responsible for the preliminary data analysis, but I was involved in deciding what that data should look like and how it should be produced.

When AT&T and Lucent split in 1996, the Labs split as well, but I think AT&T didn't take much of the physics area, so it was less obvious to us. The math department was split right down the middle. That was tough. We stayed with Lucent.

Lucent offered me a package to retire, which I took in 2001. I have remained professionally active, but I'm winding down now. I have been trying to keep up with a certain corner of the *Ulysses* data that no one else ever looks at. There was a meeting in Greece in May 2008 I wanted to attend, so I worked up some of that data and presented it. Lucent let me keep my office until 2008. It was something to get that cleaned out. There was probably forty years of stuff in there! I moved some of it into Lanzerotti's office. He's mainly at NJIT these days, but still has an office at Lucent.

Lanzerotti has been a very generous collaborator. He always gave credit to those he worked with. We worked together very well. He was very good at seeing projects and thinking of things to do. With a little bit of a push I would go on that track and get those things done. My chore was getting into the computer and pulling out the data, and seeing things in the data. I may not always know the physics behind it, but if we have been looking at something for months and it looks like this and all of a sudden it looks like something else, I can pull that out and maybe figure out why. Lou and I worked together a long time.

Looking Back

Looking back, my career at Bell Labs was very exciting. I think today I would have been pushed a little harder to get an advanced degree. I'm not sure an advanced degree in math would have been the right thing, and I probably would have gone into computer science instead. At one point in the 1960s I started taking some graduate courses in physics at NYU. The Labs supported that by giving me time off and by paying the tuition. However I talked to Walter Brown at some point and he said, "This is not what is going to get you promoted." He said, "We expect you to learn what you need to know on the job." He saw that my position was not doing the physics; that someone else would do that. I was more valuable to them doing what I was doing. I learned enough of the physics to be useful. In 1967 I bought a house and stopped taking the graduate courses.

Once I started working in the physics area at the Labs, the people I worked with recognized my strengths and I got to work on a lot of different projects and meet a lot of interesting people. The *Ulysses* team had members from France, Greece and Portugal as well as Berkeley and Kansas; at our team meetings in these places we always tried to absorb some of the local flavor. I've done a lot of interesting things and been to many unique places.

Once we stopped in Thule, Greenland on our way to the scientific radar in Sondrestromfjord. It was a military plane and we had to catch it at McGuire Air Force Base. Another time we stopped in Goose Bay because it was too early and we were not allowed to land that early in Greenland. So we stopped at Goose Bay for a couple of hours in the middle of the night. In the 1990s I gave an invited paper at a meeting in Longyearbyen, Svalbard, in the north of Norway. One day during the meeting a few people were going further north to look at a rocket launch site; they had an extra place in the plane and asked if I'd like to go along. The launch site was near Ny Alesund, which is where the dirigible was launched that went to the North Pole in 1926 with Nobile. It was about 78 degrees north. So I have been to 78 degrees North and 90 degrees South! I have had a terrific career!

ALAN G. CHYNOWETH

Alan Chynoweth. (Reprinted with permission of American Institute of Physics)

Introduction: To a Shining Laboratory on a Hill

The path that led me to Bell Labs, and the career that I had there, had many unexpected twists and turns. I picture it as a sort of Brownian motion reflecting many unanticipated coincidences or highly fortuitous breaks. But as I look back on it, I am aware of the debt I owe to so many people who gave me those breaks or influenced me at key moments and over various segments of that path.

I suppose the path really started in my boyhood and, no doubt, the genes passed on by my parents had something to do with it. My father, who had been educated as a mechanical engineer, was responsible for the maintenance of a large hospital on an estate of some six hundred acres. The hospital was relatively self-supporting with its own power station, waterworks, and farm. Our home was associated with the farm on this estate, set in the extraordinarily beautiful countryside of southern England and just an easy cycle ride, three or four miles, to the creeks that opened into the English Channel. Tellingly, engineering — especially that associated with mining — is the major theme running through my paternal ancestry, which traces back through many generations involved with copper and tin mining in Celtic Cornwall. "Chynoweth" is a Cornish name, meaning "New House." In "Visitations by the College of Heralds," aimed at verifying the legitimacy of our coats of arms, a family tree shows the origin of the name in the 1400's, in the reign of King Edward IV, with the new house being built in the parish of St. Erth, not far from Penzance. This location is noted for mining activity stretching back to prehistoric times.

Where my mother's side traces back to is much less certain, but her maiden name, Fairhurst, appears to suggest Scandinavian roots. From genealogical researches it is believed that the name is descended originally from Boernicia, a kingdom populated by Scots and Angles in the north of England, along the Scottish border, an area quite likely subjected to raids and settlement by those great seafarers and explorers, the Vikings.

Could it be that those twin threads – the Celts involvement with mining, metals, and engineering, and the Vikings seafaring adventurousness – find echoes in my lifelong enthusiasms for applied science on the one hand and globetrotting on the other?

Even if this is a bit of a reach, my boyhood years provided plenty of stimuli for these interests. I always jumped at the chance to visit "father's" engine room, to see those mighty diesel engines (of early design) driving the electric generators, the water pumps gushing to fill the large tanks in the tower above, the huge coal-burning Lancashire boilers tended by sweating stokers, in a seemingly dreadful inferno, to provide heat and hot water to the hospital. And, in a quiet side room, I was intrigued to watch the stepping switches of an automatic 100-line telephone exchange operate as calls were being placed. (This was in the mid-thirties; when we first arrived in New Jersey in 1953 I was amazed to find that, almost in the shadow of Bell Labs, there was no dial on our apartment's telephone; to place a call we had to lift the receiver and wait for an operator to come on the line!)

Then there was the lust for travel. I especially loved trains, watching them as well as taking them to (seemingly) far off places. And ships. We lived nearby to Portsmouth and Southampton and it was always a thrill to see the warships and liners. How they stirred my imagination, trying to visualize the faraway exotic places that they went to over the sea.

Even as a very young boy, I remember being fascinated by the occasional dramatized accounts of scientific discoveries and about the scientists who made them, which were broadcast by the BBC in "Children's Hour," a program that aired over the radio (in those pre-television days) while the family was gathered around the table at tea-time. The achievements of scientists such as Isaac Newton, Huygens, Davy, Galileo, and Faraday are some that I can recall hearing about.

Perhaps the greatest stimulation, however, was World War II. Starting in September 1939, and lasting for six years, one was exposed to a steady stream of news about developments in military technology, on the one hand, and about military campaigns in many different parts of the world on the other. Where we lived, on the south coast of England, we had a ringside seat for some of the war action. In 1940 we watched aerial dogfights above us, taking a particular interest in the almost daily activities of "our" local airfield, adjacent to our farm, on which were stationed Spitfire and Hurricane fighter squadrons. Then later, as the preparations for D-Day built up, we found ourselves living in the midst of a vast armed camp; we could see military vehicles, tanks, guns and troops, everywhere we cycled, camped in the fields or hidden in woods. And on the creeks nearby, a huge number of landing craft were assembled together with some weird looking contraptions whose intended use we were puzzled by; they turned out to be pieces for the artificial Mulberry Harbors to be established on the invasion shores. It was all heady stuff for a young teenager. When the atom bombs ended the war, I remarked to some farmers with whom I was working during my summer vacation that "next they will be putting atom bombs on rockets." It seemed obvious to me, after our experiences with German "doodlebugs" and rockets, but they were horrified at the suggestion and regarded me as a menace! And I was just as surprised by their reaction.

At high school, my favorite subjects, naturally, were science and mathematics on the one hand, and geography, learning about the world, on the other. But it was physics, seen as a key to understanding many technologies, that captured most of my attention. So, having passed the necessary examinations, in 1945 I went up to London University King's College to "read" physics. The department was just recovering from wartime depredations and there was only an acting head. About two months into the first term, we saw him approached by two men whom we took to be visitors. That evening, on the BBC news, we learned that he had been arrested as an "atomic spy" (the first). Subsequently, the College hired a new professor as head of department, John Turton Randall; it would turn out that he would have quite an influence on my subsequent career.

I enjoyed all the undergraduate courses but Randall's lectures intrigued me the most. He was a very low-key lecturer, not at all the sort one would expect to inspire students. His lectures were on "Modern Properties of Matter" yet, for some reason, I was "turned on" by such topics as crystal structures, band structures, and the mechanisms of various types of luminescence. Perhaps it was because before the war Randall, along with his colleague Maurice Wilkins, had worked with the General Electric Company (G.E.C.) research laboratories in Wembley on luminescence in various phosphors, work that led to a line of luminescent lamps manufactured by G.E.C. It was an example of the development of scientific understanding leading to practical applications, a theme that I found very satisfying.

The physics department at King's was named after its founder, Wheatstone, and his original "Bridge" was on display in a cabinet. Wheatstone himself was a quite prolific inventor and among his other inventions on display were early electric telegraph devices. I also heard about a more modern possibility for telecommunications. A fellow student in the physics department was Arthur C. Clarke, who later became famous as a science fiction writer. But I recall hearing him give a talk at one of our afterhours departmental Maxwell Society meetings (James Clark Maxwell did much of his seminal work while head of the physics department at King's). His topic was based on an article he had just published in a radio journal, *Wireless World*, in which he proposed that rockets could one day place telecommunications relay satellites in stationary earth orbit. I can still visualize him drawing the illustrative diagrams on the board.

For living quarters while at college I was extremely lucky to be accepted for one of the very few places in a university hostel, Connaught Hall. (Most students had to live in private lodgings.) It was an eclectic lot of students who stayed there, about sixty in all. They covered a wide range of disciplines, there were both undergraduates and graduate students, and, between them, I remember counting seventeen different countries represented. At one time we even had a monk from Tibet! Interactions and conversations among the residents could only stimulate my curiosity about other fields of study and also whet my appetite to see other countries in future. Looking back on it, my four years in the Hall were an extraordinarily enriching experience.

I had originally expected to go in for teaching when I got my degree but meanwhile, chatting one time with one of the graduate students in Connaught Hall who was doing research in some area of physics, my eyes were opened to a possible career path that I had not been aware of before. It seemed like it would be fun. Fortune smiled. On completing the undergraduate degree, I was invited by Professor Randall to embark upon research. There were various fields of study available in the department: nuclear physics, theoretical physics, radio and atmospheric physics, and a new-fangled area that Randall was initiating with his former colleague, Maurice Wilkins, called bio-physics. I chose nuclear physics. When Randall asked me why I did not choose his area, I replied that, with the enormous national interest in atomic energy, I felt that it was more likely to offer job prospects afterwards. In retrospect, it was an unwise decision since the resources available for doing leading edge nuclear physics research were not really adequate. I embarked on assembling instrumentation for studying nuclear radiation spectroscopy but, by a stroke of sheer luck, Randall received and brought to my attention a thesis publication he had just received from a scientist in Holland: it was on the use of electrically insulating crystals such as some alkali halides and even diamonds as induced conductivity nuclear radiation detectors. Consequently, my research veered into a study of these potential devices.

Our relatively small nuclear physics group was housed in a somewhat isolated area, mainly in subterranean cellars. There was a variety of work, including quite spectacular cloud chamber studies of nuclear reactions and photographic plate studies of cosmic ray showers. One very talented student, who became a close friend, was Raja Ramanna from India. On completion of his doctoral work he returned to India where he achieved considerable fame and recognition, coming to be regarded as the "father" of India's atom bomb and eventually serving as Minister of Defense.

One of the laboratory assistants who served the group was a young woman who had arrived in England just before World War II as a refugee from Austria. Her surname was Kompfner. One day she brought in a visitor, her brother, Rudolph, who, I believe, was working at the Clarendon Physics Laboratory in Oxford University. Supported in this work by the Admiralty, he was credited with inventing the traveling wave tube microwave amplifier. He showed a delightful, enthusiastic interest in all the work going on in the group. I could not have guessed that our paths would converge again one day.

As a result of Randall's intervention, my work veered away from nuclear physics and more towards the field that later became known as solid state physics. I set out to study crystals as an alternative to Geiger counters for monitoring radiation. Fortunately, the College's professor of geology, William T. Gordon, was a recognized authority on diamonds; he spent many a summer vacation grubbing around in African diamond fields and had amassed an extremely impressive collection of natural diamonds in his filing cabinet.

He happily allowed me to rummage through and select half-a-dozen for my studies. They were beautiful, large, clear specimens shaped as slices through an octahedron, some more than a centimeter across and three or four millimeters thick, just perfect for mounting electrodes. Gordon was perfectly willing to lend them to me, but he made a careful note of their catalogue numbers and of my borrowing them!

It was the frequent custom for members of the department to congregate in the afternoons in the departmental "tea-room". There we had the opportunity to discuss our work informally with any interested colleagues and faculty. There were also journal racks. One day, scanning through the *Physical Review*, there was a short article on bombardment induced conductivity in diamonds authored by someone at a place called Bell Telephone Laboratories. (I remember thinking it was a strange topic for a telephone company!) Maurice Wilkins, who later shared the Nobel Prize with Crick and Watson for unraveling the double helix structure of DNA, was particularly helpful to me thanks to his previous experience with luminescence phenomena and encouraged me to submit a short article on the results I had been getting, which I did.

I also remember the stir among those in the tea-room when an article appeared in the *Physical Review*, also from Bell Telephone Laboratories, announcing the observation of a voltage or signal amplification effect in a semiconductor crystal portending a solid state device, for which the name "transistor" was proposed. The significance of this news – of a potential miniature, low-power replacement for triode vacuum tubes – was not lost on my more senior colleagues.

As my thesis work seemed to be proceeding quite satisfactorily towards its conclusion, I started to think about the next step in my career. I particularly felt it would be a good time to try to see some of the world before settling down in Britain. I scoured the advertisement pages in the scientific journals with a view to seeing if there was anything attractive on offer in other countries, particularly English-speaking ones. I was attracted by an announcement by the National Research Council (NRC) of Canada of the availability of post-doctoral research fellowships in various fields of research. I applied, stating my preference, in order, for three of the several fields offered: i) Nuclear Physics (at Chalk River), ii) Solid State Physics, and iii) Physical Chemistry, these latter two stationed at the headquarters laboratories in Ottawa. To my surprise, I received an offer – in Physical Chemistry! Of course, with the youthful confidence that a physicist ought to be able to turn his hand to almost any branch of science, I accepted. When word got out around the department, I remember one of my biophysics colleagues, the ebullient Raymond Gosling, greeting me with the remark, "I hear MacKenzie-King has sent for you!" (M-K was Canada's famous wartime and immediate post-war prime minister). Looking back on my thesis work, it had the result of starting a new field of research for the physics department: studying the electronic and optical properties of diamond, both natural and synthetic, with particular emphasis on the roles of impurities and imperfections in what is, after all, the most extreme member of the semiconductor series, germanium-silicon-carbon. This program has now been in existence for sixty years.

So, with my new bride, Betty, who I had known since high school, we sailed for Canada in October, 1950. When I arrived at the NRC, I and other newly-arriving post-docs were greeted by the genial head of the Chemistry Division, E. W. R. Steacie. He expressed the hope that we would make good use of the fine facilities at NRC to further our scientific careers but he also urged us to find time to get to know Canada, to "try to acquire a car and see something of the country." (In retrospect, perhaps Lady Luck was shining on me yet again. The joke among the post-docs was that those arriving in the Physics Division, being greeted by G. Herzberg, were told that "first, they should get a pass so that they could work in the labs on weekends!"). We took Steacie's advice. With Betty helping by earning some money by becoming a secretary, we were able to buy a car and with it we explored a lot of territory in Ontario, Quebec, and across the border, in New England. Towards the end of my two-year appointment, we even took off for three weeks and drove out to the West Coast and back via the northern States and the Rockies.

The physical chemistry group was headed by W. G. Schneider. I was thankful that the work was more physics than chemistry. My main line of work was a study, by ultrasonic interferometer methods, of critical point phase transitions in gases and liquids. Schneider also was curious about how organic compounds in the eye's retina converted incoming light into electrical signals. He suggested studying photo-effects in such molecular materials as beta-carotene but I felt it would be better to start by picking something physically simpler to work on, namely, readily-available single crystals of anthracene. I was able to obtain results on photoconductivity in these crystals which suggested that they behaved, conductivity-wise, in some ways similar to inorganic insulating crystals, such as the diamonds I had done my thesis work on.

One day, Schneider drew my attention to a meeting of the American Physical Society that was to be held at the G. E. Laboratories in Schenectady, New York and suggested that I might submit an abstract on the diamond crystal work I did in London. Consequently, I gave my short talk at the meeting and afterwards a gentleman came up and introduced himself to me. (Lady Luck again!) He was J. B. Johnson (of Johnson-Nyquist Noise fame) from Bell Telephone Laboratories. He told me of some of the related work on diamond conductivity that was going on in the physics research group at Murray Hill and invited me to visit them. A few weeks later, I traveled by overnight train from Ottawa to New York and made my way to the old West Street Laboratories. From there, a ride on the company car service took me to Murray Hill where I was met by K. G. McKay, my host for the visit.

What an eye-opener! For the rest of that day and much of the next I enjoyed a series of visits to members of the physics research group that was headed at the time by A. H. White. I was fascinated by the variety and quality of the work that was described to me. Besides McKay's studies of bombardment-induced conductivity in diamond, there were investigations under way of surface physics, thermionic cathodes, breakdown in gases, theoretical physics, electrical breakdown in semiconductors, and secondary electron emission, to name just a few. I was struck by the whole ambiance of the laboratories, the sophistication of the equipment, and perhaps most importantly, the enthusiasm of the members of staff and the courtesy that they and the managers showed me. It really did seem to be "a shining laboratory on a hill." On taking my leave at the end of the visit, I enquired of McKay what possibility there might be of my joining the Laboratories at the end of my two-year post-doctoral appointment in Ottawa since I desired to round out my North American research experience with two or three years in the United States before returning to

Britain. In reply, it was suggested that I make another visit so as to explore the possibilities more broadly.

A second visit was duly arranged. I arrived at Bell Labs not quite knowing what to expect. I was naive. I should have realized that I would be asked to give an account of my research work, but I came totally unprepared for this initial step in the assessment ritual. It came as a complete surprise to be ushered into a room to find a number of researchers waiting to hear what I had to say! Without slides or props of any kind, all I could do was ad lib at the blackboard. Afterwards came a series of interviews in which I learned more about what was going on while, at the same time, the interviewers would be forming their own impressions of me. It was a procedure that I was to become very familiar with in later years.

The year was 1951. Excitement was in the air at the Labs. The transistor had recently been demonstrated, not only with point contact structures but with grown p-n junctions in germanium and silicon crystals. There was the feeling that a whole new field had been opened up in which exploration of the properties of single crystal materials might yield a cornucopia of new phenomena and potential applications. Clearly, my timing was fortuitous as there must have been a real hiring effort underway to build up the relevant research programs. The interviews revealed many more exciting studies in progress, especially exploring the properties of semiconductors. One of the interviews

was with a couple of organic chemists who were particularly interested in my work on photoconductivity
in anthracene; in retrospect, I now understand why
knowing looks were exchanged between them since
one of them was Bill Baker.

I suppose candidates from abroad were still somewhat of a novelty because, apparently, the Director of
Research, Jim Fisk, and the Vice President for Technology, Ralph Bown, had asked to meet me. I was
ushered into their presence. After the usual courtesies,
Fisk, pulling on his cigar and noting that I was from
the physics department of London University King's
College, started to enquire as to the professor there:
"Was it X?" No, I replied. "Oh! Was it Y?" No again. I
thought this was proving to be quite an easy interview
as Fisk seemed to be enjoying his guessing game, so I
did not interrupt it. Bown looked on with an amused
smile. Eventually, Fisk gave up but when I told him
it was Professor Randall he exclaimed loudly, "Of
course!" Apparently, during World War II, Randall had
delivered to Fisk the prototype of the cavity magnetron
that he, together with colleagues Boot and Sayers, had
invented and which played such an important role in
the Allies' conduct of the war. Fisk had been involved
in developing the magnetrons and getting them into
production at Western Electric. After chatting some
more, Bown suggested that while I was visiting I might
like to see some of the interesting work going on in
the device development area. As tactfully as I could, I

declined the suggestion as at that time, as an ignorant foreigner, I had only a rather negative image of what the atmosphere might be like in an American industrial development division. Perhaps Hollywood was to blame but later in my career I was to find out just how exciting the Bell Labs Development Area could be.

Over lunch, the conversation at one point turned to describing some of the benefits provided to employees. I was told by McKay, perhaps a little apologetically, that the vacation allowance was only two weeks but that this could sometimes be stretched a bit. For example, Walter Brattain had done some nice work recently (co-inventing the transistor) and so the company allowed him to go on a two-week lecture tour, visiting universities around the country. Times have changed!

I was certainly enjoying my visit and the interviews until I was brought up short by a session with W. Shockley. He had just completed his book *Electrons and Holes in Semiconductors,* an extraordinary piece of work with chapters that generally alternated between fundamental properties of semiconductors on the one hand and possible electrical circuit applications of transistors on the other. Each chapter ended with a set of problems "for the student." Shockley's interviewing technique was to ask me to work out some of these problems at the board! After this embarrassing ordeal, I felt there was little likelihood of my getting an offer of employment and I returned to Ottawa in despondent mood.

On returning to NRC, I immediately visited D. K. C. McDonald, who had recently been hired into the physics division from Oxford University to build up a low temperature physics group. He indicated he could offer me a junior scientist position but, shortly afterwards and to my enormous surprise, a letter arrived from Bell Labs offering me a position in McKay's new group. So, after an extraordinary, Kafkaesque experience with US government bureaucracy, which needed Bell Labs' lawyers to sort out (a story in itself), I became a member of the technical staff (MTS) in the physics research division at Murray Hill on March 3, 1953.

At the Shining Laboratory on a Hill

The Laboratories at Murray Hill. (Reprinted with permission of Alcatel-Lucent USA Inc.)

The day after I started, who should appear at the door of my laboratory to welcome me to Bell Labs but Rudy Kompfner. A small world! Since my meeting him at King's he had been invited by fellow inventor John Pierce to join Bell. Rudy, with that wonderfully friendly manner of his, quickly helped to make me feel at ease in my new surroundings.

As an MTS in physics research during the fifties, there was never a dull moment. Lunch conversations, seminars, and interactions with various colleagues all provided an outpouring of new discoveries, phenomena, ideas, and inventions, on an almost daily basis. Looking back, I often think of the 1950s as the halcyon days of solid-state physics. And Bell Labs was a wonderful place to be doing such research. As was often realized, we had all the benefits of academic freedom, along with good resources, and none of the teaching or administrative loads that our counterparts in academia usually faced. Furthermore, compared to academia at that time, the pay was relatively good. After I had been there for a couple of years, McKay asked me whether I was still intending to return to England, but Bell Labs by then had seduced me! I chose to stay if they would have me.

Perhaps naively, at the working level we were hardly aware of "management," especially at the lofty strata above our immediate supervisor. It was clear to me in later years, however, that at that time it must first have been managers such as Ralph Bown, Jim

Fisk, Bill Baker, the president Mervin Kelly, and others that had the vision and wisdom to ensure the support of this creative environment. Further, I hardly noticed whatever changes took place in the ranks among senior management but at some point in the 1950s, Bill Baker became Vice President, Research. Then one day I realized, much to my amazement, that Bill had become aware of me. It was when I found myself pushing my tray along in the cafeteria right behind Bill. He turned to me and remarked: "Alan, that's very nice work you are doing on the emission of light from switching in ferroelectric crystals!" He got it half right. At the time I was pursuing two separate lines of research: light emission from avalanche breakdown in silicon p-n junctions and the mechanism of ferroelectric domain switching in barium titanate crystals. I murmured appreciatively, feeling it prudent not to correct him. But that aside, Bill was renowned for his prodigious memory, not least of all for remembering the names of countless employees and what they were doing even though he seldom met them.

But for most people, most of the time, Bill seemed a rather remote, even mysterious person with whom those at the lower levels communicated, if needs be, through the interposed layers of management. Likewise, when Bill's views came back, it often seemed that those middle managers relayed the messages in an almost hushed, or reverent, tone as if they had just been in touch with a deity. Many a time those managers also had another task to perform, namely, translating what Bill had said or meant. Bill's complex prose was legendary. When the need arose, he could be very clear and direct but most of the time he seemed to take great delight in being enigmatic, even ambiguous. Perhaps he sometimes did it deliberately in order to stimulate others to think for themselves. If so, he certainly succeeded. Deciphering his messages became a teasing and frequently amusing challenge for many of us.

After a while I had established some quite fruitful lines of research, which brought me into contact with a growing number of colleagues, not only in research but in the device development area as well. These interactions steadily broadened my interests as well as making my work seem all the more relevant to possible applications (that combination of basic understanding and its applications that I had originally recognized as an important "turn-on" for me). The breadth of my interests started outstripping what I could effectively encompass in my own research and so, though I had never contemplated becoming a research manager, perhaps it was inevitable that eventually, and to my surprise, I was invited to become a research department head with responsibility for a quite small but diverse range of research programs.

Many immigrants from time to time experience feelings of homesickness. This was especially true, I can personally attest, for non-working spouses. And the uncomfortable and unaccustomed heat and humidity of typical New Jersey summers in those pre-air conditioner days, compared to more temperate climates, only served to intensify the feeling. Thus, after three years in America, Betty and our two-year old son sailed back to England for a summer vacation. I, of course, with a measly two-week vacation allowance, was hardly in a position to accompany her in those days before air travel became feasible. But after another two years passed, I had carried over some vacation so that I could afford to make the transatlantic trip. My boss at the time, Gerald Pearson, generously suggested I might extend my vacation by visiting some industrial research laboratories in England and on the continent. Of course, I was pleased to follow up the suggestion and so was able to combine my interest in traveling to other countries with some work-related purpose.

In the coming years I was to have many more opportunities for foreign travel. Some of these were to International Semiconductor Conferences in Prague, England, Paris, Moscow, and Japan. But another purpose arose for making regular visits to Europe. Recruiting of new members of technical staff was usually done via knowledgeable scientists and engineers visiting the universities from which they had graduated. But Bell Labs also frequently received enquiries about employment prospects directly by way of letters from abroad, from science and engineering graduates or from their professors writing on their behalf. Naturally, such letters would make it seem that the applicant could "walk on water." And since there seemed to be no substitute for screening the candidates with face-to-face meetings, the notion got raised that I, sometimes together with Klaus Bowers, should make trips to Europe twice a year to follow up on these letters and interview the candidates and relevant professors. These trips regularly took us to such places as Oxford, Cambridge, Imperial College London, and ETH in Zurich. Other universities were visited as the need arose. These recruiting trips lasted through the sixties and resulted in over one hundred graduates being added to Bell Labs' technical staff. It was a tough job but someone had to do it! There were many offers of help from colleagues but we willingly made the sacrifice. This recruiting activity, which was undoubtedly blessed by Bill Baker, certainly brought an enriching diversity of knowledge, backgrounds and perspectives to the research and development organizations. I even heard my own department referred to as "Chynoweth's foreign legion!" However, by the late 1960s, as Europe steadily recovered from the war, more and more opportunities were opening up closer to home while, at the same time, the Viet Nam war was having a negative effect on America's image. Consequently this source of new employees tended to diminish.

After a few years as a physics research department head another invitation came – as a complete surprise – namely, to become a co-director, with Jack Scaff, of the Metallurgical Research Laboratory. What did I know about metallurgy? I couldn't help but think of that patter song out of the Gilbert and Sullivan operetta, *HMS Pinafore*, in which Sir Joseph Porter, KCB, proclaims, "He had never seen a ship and never been to sea, so they made him the ruler of the Queen's navee!" (Was it in any way pre-ordained by those Cornish genes?) I was intrigued and accepted. Some of my physics colleagues raised their eyebrows at my transferring to 'The Forge,' as they rather disparagingly nicknamed it, perhaps reflecting the traditional pecking order one often finds on the university campus.

In this position I was succeeding another physicist, Al Clogston. Clearly, the opportunity was there to continue bringing a physicist's modern perspective to the venerable field of metallurgy – or so I thought. I was soon disabused of that thinking when I was introduced to the wide range of exciting activities going on in this relatively large organization. Besides the expected physical metallurgical work on various metals and alloys, including magnetics and superconductors, there were fundamental studies on crystal growth, especially magnetic garnet crystals, piezoelectrics, and various crystals for possible use in optical communications devices, on thin films, glasses, and ceramics. There were metallurgists in the organization, of course, but members of other disciplines were strongly represented as well, including physicists, inorganic chemists, ceramists, and some mechanical engineers.

It was truly a large, interdisciplinary laboratory. And besides the range of fundamental research, there was nearly as much work going on in development work, in which there was close cooperation not only with development areas within Bell Labs but with Western Electric manufacturing locations as well. It was a truly rich and exciting mix of activities ranging from science to what could only be described as art.

The example of the latter that stands out in my memory involved a researcher who was attempting to find a suitable combination of oxides for making ceramic substrates with various critical properties essential to supporting hybrid integrated circuits; it was pure witchcraft as far as I could tell, relying entirely on an empirical approach and the accumulated experience of the researcher. Physics had nothing to offer to this sort of project.

Despite what seemed to me to be an extraordinarily important range of work going on, I discovered that morale in at least parts of the Metallurgical Laboratory was quite low. There was the feeling that they were regarded less favorably, budget-wise, by upper management and got less publicity than the glamorous activities of the Physics Research Laboratories. I saw no reason for such an inferiority complex. Apparently, at one time they had sought to get the name of the laboratory changed but Bill Baker had turned down their request. I did not understand this and took it upon myself to give it another try. I proposed that the name should be changed to "Materials Research Laboratory," arguing that this term was a more appropriate fit to our activities. Bill immediately approved it! Maybe my timing of the request was fortunate, but it certainly anticipated other events yet to come.

Besides the Metallurgical Research Laboratory there was the separate and large Chemical Research Laboratory, focused mainly on polymers and plastics, very much Bill's original area of expertise and responsibility. Together, these two laboratories made up the Materials Division and perhaps a possible confusion of organizational names explains why Bill was initially against changing the Metallurgical Laboratory's name. There was a friendly rivalry between the two laboratories, particularly regarding whether metals or plastics would be best for some applications in the Bell System, or whether our work on glass fibers for optical communications, for example, was more important than the chemist's work on anti-oxidants for stabilizing cable sheaths. Of course, some of these rivalries would be aired in the annual merit and salary competitions conducted by our mutual boss, Bruce Hannay. On one occasion, my counterpart in chemistry, Bill Slichter, trying to score some points, remarked that they had a few modest salary adjustments to propose. That triggered me to exclaim, recalling Winston Churchill's famous comment on Clement Attlee, "They have much to be modest about!"

In retrospect, I think Bill Baker must have had a soft spot for the work of the Materials Division, not only because it was close to his own technical expertise but because it gave him a steady stream of contributions to the technology of the Bell System, which he could use to advantage in his negotiations and other interactions with our parent organizations, AT&T and Western Electric. (I suspect it also helped him demonstrate repeatedly the value of research to his fellow officers in Bell Labs, some of whom were possibly inclined to claim that their areas were more than capable of managing without "help" from research!) But there was also another motive at work. Bill recognized, and was a profound believer in, the power of mission-oriented interdisciplinary research, which the materials field exemplified so abundantly. At the same time, his connections in Washington, in the National Academies of Science and Engineering, and in various government advisory committees, gave him opportunities to promote this view with evangelistic zeal. In particular, he was a prime mover in urging the formation and support of interdisciplinary materials science centers in universities. This was at a time when the word "interdisciplinary," let alone the idea of mission- or problem-oriented research, seemed often to be anathema to universities organized along traditional disciplinary lines. Their bias was against working on anything that smacked of applied research. After all, this type of work could put a young faculty person at quite a disadvantage if tenure was being sought! Yet I believe that applied research is often more challenging than so-called pure, or basic, research. In the latter, the researcher can generally pick idealized models or set the boundaries to the research in such a way as to provide a reasonable chance of finding an answer in due course. In applied, problem-oriented research, the desired goal is more or less set for the researcher at the outset and the challenge is to find a way of reaching it, usually on a definite time-scale and with only sketchy prior knowledge of the fundamentals or underlying science. In the field of materials research, and even more so later on when I became responsible for major electronic and photonic device development programs, I was to have numerous occasions in which this observation got reinforced.

A major example of this, which I remember well, was when AT&T Long Lines, in association with its counterparts in Britain and France, decided to put a transatlantic optical cable into service some eight years hence even though, at the time, we were not at all sure that we would be able to develop the optical devices (semiconductor lasers) with the excruciatingly high degree of reliability required. In my mind I likened it somewhat to President Kennedy's announcement of going to the moon in ten years' time even when there were so many unknowns about the necessary underlying technologies.

I was to have another and totally unexpected opportunity to help promote the image and value of the interdisciplinary materials research field. From time to time, the National Academy of Science supported studies of the status and potential of the major disciplines of science – particularly physics and chemistry. I do not know what discussions led to a study of the materials field, but I am sure Bill played an important role. As a result, a study of the field of materials science and engineering (MSE) was commissioned under the auspices of the National Research Council, an operative arm of the science and engineering academies. It was called the Committee on the Survey of Materials Science and Engineering, COSMAT for short. The highly esteemed professor of metallurgy at M.I.T., Morris Cohen, was appointed chairman of the survey and Bill was appointed as co-chairman. To assist in the sizeable amount of work entailed, Morris brought in one of his metallurgical colleagues but Bill quickly realized that, in spite of all the good will in the world, the study risked over-emphasizing the metallurgical field to the detriment of the far wider field of materials that Bill had in mind. I was therefore brought in to help with the study. There followed an intensive couple of years of work, organizing panels, soliciting input from leaders in academia, industry, and government, and writing reports. It was an extraordinary exposure to the technological heart of the U.S. economy.

One of the first tasks that faced the committee was to figure out who worked in, or which disciplines comprised, this field of MSE. We solicited views from various research organizations and professional societies by means of surveys. I recall the surprise with which one well-known metallurgist greeted the results of the survey: "Even the solid state physicists see themselves as working in the field of MSE!" And together with ceramists, polymer and plastics chemists, and many branches of the engineering disciplines, it was clear that the field of MSE was, indeed, interdisciplinary. The overall Survey went on to outline what Bill described as a veritable blueprint for America's technology. Morris Cohen, armed with the imprimatur of an Academy study, went on to proselytize the message about the value of the interdisciplinary field of MSE in academia and the professional world in general. The Survey may well have given added impetus to the support of the Materials Science Centers at universities. It also saw the birth of the Materials Research Society, largely through the efforts of Survey member Professor Rustum Roy of Penn State University. This has become a leading professional society with an impressive meetings calendar and steady output of technical papers.

If the materials field was not broad enough for me, there were always the richly diverse activities of the Bell Labs research itself. Many of these we first got to hear about at Bill's monthly three-level staff meetings. Crass administrative matters only rarely intruded on these meetings. They were much more opportunities for the twenty or so directors and executive directors responsible for the various laboratories making up the research area (including John Pierce and Rudy Kompfner) to "show and tell." We made sure we were always in place waiting for Bill to sweep in, utter his enthusiastic greeting – "Good Morning, Gentlemen!" – and start the proceedings with some fresh and often fascinating news or insights from his current experiences in Washington or, sometimes, from developments with our parent company, AT&T. (On one occasion he was giving us some news from the White House when his secretary came in and whispered to him; he left the room quickly saying, "Speak of the devil!") After Bill's opening remarks, the rest of the morning would be taken up going round the table. To me, it was always stimulating to hear about the new discoveries, results, theories, ideas and inventions emanating from the various centers, ranging from the physical sciences to communications technologies, software to economic theory, mathematics to behavioral sciences, chemistry and even patents. It was nearly always leading-edge stuff. And Bill had a remarkable ability for moving the meeting along in good-humored fashion and at the same time, showing knowledgeable appreciation and having something nice to say about each presenter's contribution.

Although we all took the show-and-tells seriously, there were often lighter moments, even when we had to spend a few minutes on administrative matters.

On one occasion, we were all given print-outs of the capital equipment on our books to help us identify items that should be junked. As I looked at my list I blurted out in surprise at finding there was a rolling mill with an acquisition date of 1911! Bill leaned over, remarking in his soothing voice, "Perhaps you should donate it to the Smithsonian!" On another occasion, Bill's administrative assistant, "Pat" Keenan, was reviewing his estimates of how many people we would be able to hire the following year, including how many people would have to be replaced because of transfers, resignations, and retirements. He also said he allowed for one death, at which John Pierce piped up, no doubt reflecting the concern we always had for our personnel, "Has he been told yet?"

Eventually, after twenty-two years in the Research Area, I got another surprise. I was invited to take a position in the Development Area as Executive Director of the Electronic and Photonic Devices Division, responsible for the development of essentially all devices intended for use in the Bell System except silicon integrated devices, those being the responsibility of two other, parallel divisions. This Division was a very good fit for my interests and background, as it was a rich mix of device physics and materials technology embracing a wide range of products either introduced, or about to be introduced, into production lines at more than half-a-dozen Western Electric manufacturing locations. But it meant I was no longer directly in Bill's domain though I remember one senior colleague remarking to me at the time, somewhat darkly I thought, that he could "see some plan afoot." I had no idea what he meant but I deemed it better not to seek clarification. Now I reported to Vice President John Mayo, another person for whom I had great respect and who, in the following several years, gave me great support and opportunities. During this period Bill became President of Bell Labs and I experienced an event which gave me a personal and very touching example of his concern for his people. After about four years in the new position, I had a sudden angina attack which resulted in my having by-pass surgery two weeks later. To my amazement, while I lay in hospital, I received a beautiful letter of encouragement from Bill. He had taken the trouble to find time in his back-breaking schedule to compose quite a long letter, which I do not need to go into, but I couldn't help but read with a chuckle his little dig about how, given time for nature to do its magic, it would be those organic chemical molecules that would get me mended.

Recalling this anecdote brings me to make a few remarks about how I saw Bill as a person. Bill was a most unusual executive, a complete antithesis of how some may think a person in such a position behaves. In his Bell Labs office he was no clean desk freak; far from it. Visiting him there one time, I thought he was not there until I heard a voice from behind the enormous piles of papers and reports on his desk. And when he went home from work, he always had one or two bulging brief cases. He lived about five miles from the Murray Hill Labs and for many, many years commuted in his ancient and rather shabby car, a Pontiac I think it was. One day upon leaving the building he discovered his car was missing. Unbelievably, it had been stolen! It turned up later in downtown Newark. Bill's car was somewhat symbolic of his modesty. He lived a very quiet and unostentatious private life, which also helped to further surround him with that air of mystery. About his own achievements he was modest to a fault yet he always made sure that others got credit for theirs, sometimes embarrassingly so. But Bill reveled in his professional, and perhaps even his managerial interactions, with other intellects while drawing on his legendary knowledge and memory. He seemed to get particular pleasure out of composing challenging prose, frequently using unorthodox grammatical constructions. And, often, with a twinkle in his eye he would enjoy word-sparring with us. One instance comes to mind: We were seated together on the plane going to Washington. I was going there to manage a COSMAT panel at the Academy with Bill as my co-chair, but first Bill had to be at the White House. He asked me: "Alan, what time would be the best time for me to turn up at the panel meeting?" Now, Bell Labs subordinates were not accustomed to being put into the position of telling Bill what to do! After a brief pause, I replied: "Any time you come, Bill, will be the best time!". He thought for a moment and then said, smilingly, "That was a very elegant answer."

Phew!

Off to Bellcore

In 1982, disaster struck the Bell System with the court-ordered break-up of the company, as a result of the famous anti-trust suit. The major result was for the formation of seven regional Bell Operating Companies to be spun off from the parent AT&T. The newly formed companies decided that they needed a piece of Bell Labs to serve to support them as a consortium. I did not think it would affect me, as I had never had any interactions with operating companies. In fact, one time when I was wearing my hat as secretary to the Electronic Technology Council that John Mayo had set up, made up of executives from AT&T and Western Electric as well as Bell Labs, I suggested that, on hearing of the break-up arrangements, AT&T's technical strategy should be clear: We should push broadband optical technologies for the long distance parts of the network and for dedicated services to major customers, and push cellular radio systems to by-pass the local companies' loop plant! But I was in for another surprise in my career. Besides the more obvious engineering and software support, the companies concluded they needed in their consortium a substantial applied research organization as well, and I was invited to create it and to head it as its Vice President. I was intrigued by the challenge, which I found impossible to resist. Now I had to reverse roles and try to figure out how to compete against the AT&T strategy I had previously thought obvious. There followed an extraordinary few months during which a cross-section of suitable experts from all across Bell Labs had to be identified and given the chance to join in this new adventure. In the end, we fashioned a quite remarkable outfit which, I believe, served our new owners well. (Of course, in this belief I am probably a bit biased just as Bill may have been about his beloved Bell Labs Research Area).

With the formation of Bellcore, as the consortium came to be called, fear of breaking the anti-trust conditions caused, in effect, an Iron Curtain to descend between hitherto close colleagues. We were afraid to have any discussions with each other except in the presence of lawyers. Gradually, we learned what we could do and not do but it was a very distasteful and dissatisfying experience, to say the least. And in particular, with the splitting of Bell Labs, so ended my ties to anything going on there and to anyone working there, including Bill. I was to be at Bellcore, having yet another extraordinary, exciting, and very different type of experience till my retirement ten years later, but that is another story.

Closing Thoughts

Looking back over the years that Bill was responsible for the research area, the word that comes to mind is Camelot. What Bill created and fostered was truly unique: an eclectic range of research work of the highest quality, which opened doors for Bell Labs researchers all around the world, thereby giving us and our colleagues in the development areas early information on discoveries and developments happening elsewhere. Could something akin to a Bell Labs be created again? In other fields, even? I am asked that from time to time but I think the answer is, "not likely," since the US government, with its anti-trust laws, destroyed a unique support arrangement, that of a regulated monopoly that had an on-going mission to advance the nation's telecommunications network. This mission gave the corporate management a direction that helped focus the technical programs. The monopoly status gave management some confidence that they could plan for the long term and therefore undertake bold new engineering programs as well as some related research that might take a decade or even two before reaching telecommunications customers. In particular, the Bell System was able to operate as a vertically integrated enterprise involved in all stages of the food chain from basic research through device and systems development to network deployment and service provision. On the downside, the Bell System may have sometimes been a bit slow to install a particular technology but when it did decide to do so, and had the regulators' approval to do so, it had the muscle to move quickly.

Nowadays, perhaps only the government can act as a monopoly and perform or support mission-oriented long range research as, for example, through the Defense Advanced Research Projects Agency (DARPA), with its support of exploratory projects in both academia and industry. For most industries, including the fragmented remnants of the old Bell System, the competitive but often wastefully duplicative marketplace – though it may sharpen up product designs to more closely meet immediate customer needs or interests – forces companies to focus on the short term and incremental products to such an extent that very little room is left for the longer term research and development periods that truly revolutionary and complex engineering advances usually require. For this, most companies now have to rely mainly on the universities and hope that they will be smart enough to pick up quickly on any promising results.

But there were also other forces working against the Bell monopoly. Before the discovery of the transistor, telecommunications – focused on conveying voice messages reliably and without distortion to their intended destination across complex networks – was a rather specialized, even arcane branch of engineering. The transistor was to sow the seeds of change.

Though it led to a flowering of solid state science and technology and would eventually revolutionize telecommunications, it also, once it had evolved into integrated circuits and digital technologies, lent itself to revolutionizing almost every other area of electronic engineering, information technology, and product development. It became everyone's technology, the genie was out of the bottle. And though the food chain still operates, nowadays each step along the chain is populated with numerous competing companies specializing in that step so that a company in the next step has a multiplicity of sources of technology from which to choose. In particular, it has made it relatively easy for new competitors to develop their own versions of telecommunications devices, systems, and services. Perhaps it can be said that with the discovery of the transistor, forces were set in motion that would eventually and inevitably lead to Bell Labs losing its unique status.

I regard myself as being very lucky to have been in Bill's Research Area during what many regard as its glory days. But besides being lucky in the general sense, I am also acutely aware of, and humbled by recalling, the very many occasions on which a chance remark, a bit of advice, and genuinely friendly interest from schoolteachers, professors, colleagues, and, especially, the various bosses I had, in Canada, at Bell Labs, and later at Bellcore, influenced my career. Once, when receiving some praise from one of these bosses for some research results I had achieved, I self-deprecatingly murmured that I felt I had simply been lucky. To which he replied, "Well, Bell Labs likes lucky people."

DAVID DORSI

Duke Dorsi at work. (Reprinted with permission of Duke Dorsi.)

Note: David ("Duke") Dorsi was interviewed by Richard Q. Hofacker on February 7 and April 4, 2008. This interview was then edited into prose, retaining Dorsi's words.

Glassblower at the Labs

I'm David Dorsi, better known as Duke. All my friends always called me that. I've lived in Stirling, NJ all my life, and I worked in the Bell Labs when I got out of the high school in 1943. I worked there until 1946, when the war ended. There was a layoff, but I came back to Bell in around 1950. In all, I accumulated forty years with the Labs as a glassblower.

Working at the Labs was a lot of fun. You met a lot of very important scientists in those days. Those were the days when they were working on the Manhattan Project to build the first atom bomb. They had all these generals and colonels and lieutenants and all from all over the country. And even from England, they had some, I guess. I remember one time seeing Marshall Montgomery come in, and Bill knew all these fellows.

One of those scientists, of course, was Bill Baker. In the earlier years, before and during the war, Bill loved to hunt. Especially we would go out woodcock hunting because we had dogs. He liked the outdoors, and we'd go out here locally and once in a while get a bird, and occasionally go duck hunting. He was always down to earth and was not a pretentious fellow. He was a lot of fun. It always amazed me how many people he knew by their first name. He always made you feel important. I mean, especially to some of the young guys coming out of college, it was like, "Wow, you know the president."

I remember one day we were doing some work. Bill said to me that they were having some kind of a presentation with a dinner at the lodge at around noon or one o'clock. He said, "Well, I want you to attend." So, this new boss I had, he says, "No, you can't go. We're too busy." I said, "Well, look, you call Bill Baker. He's the one that invited me." He says, "Oh, Bill Baker invited you." He says, "Oh, you better go." "Yeah, you're damn right I better go." I really enjoyed my time at Bell. But let's start from the beginning.

Beginnings in Stirling

I was born in 1925 and went to school in Stirling. The town was mostly Italian with their own Columbus club and a big parade and fireworks on August 15th. My folks were from Italy, and I'm the youngest of ten. My father came first from Italy around 1908 or 1909, and then my mother followed. Neither of them ever went back. My father hated Italy with a passion. I tried to tell him that things had changed. Finally we had a travel agency, and I used to tell him, "Pop, you know, we could go over and see your brothers and sisters." He said, "I don't know. You go there. I know what it's like." I said, "But it's changed. Times are not like when you were there." But he remembered the feudal system, where the Dons owned the land and people were treated like slaves.

I grew up during the Depression, when everybody was poor. Back in those days there were peddlers who would bring vegetables to town, there were bakers who would deliver the bread to your door, and there was a milkman. We had our own cows and everything else.

I went to high school in Morristown and graduated in 1943. Before going to school, you'd get up early. We used to go out trapping and look and see if we could get some muskrat or mink or whatever we could catch. Then we'd come home and milk the cows and feed the chickens and then get ready to go to school. We used to milk the cows – one time, we had as many as four cows – and deliver the milk for 10¢ a quart. We washed the bottles and everything, carried the damn thing up the hill. We used to sell the eggs from the chickens, 25¢ a dozen. We got the school bus at 7:30 in the morning. You did a day's work until you went to school. They had one bus for the girls and one for the boys.

My mother used to make cheese and butter, and then we went hunting this time of the year [in early spring]. There were no deer the way there are now. We'd hunt for days down in the swamp, but there would be a lot of pheasant and stuff in this area because there was a lot of farming here. This was before the Great Swamp was a protected area.

We didn't participate in many sports because you didn't have a special bus to bring you back home, and you had to get home anyway to do your chores. It wasn't easy. But then the war came, and it changed the whole thing. From our town there were a lot of fellows who went in. I had four brothers and a sister who were in service. My brother Dan and I worked in the Labs at the time. I was deferred from military service because we were doing some type of work that got a deferment.

The streets back then were all dirt streets, and the funny part was, like now, they were so worried about the little bit of oil from your car – but that used to keep the dust down. They'd have big tankers with crude oil and waste oil and spray the street so as to keep the dust down. There was an asbestos plant here in Millington and they used to use the waste as fill all over town.

I remember right after school, before I got the job in the Labs, there were always openings at the asbestos plant in Millington. I worked, I think, one or two nights, and I'd come home and there was so much dust that would settle on the cars, the guys would have to go out and wipe their windshields. I could hardly breathe. My father said, "You, hell no, can't take that job," so I quit. A good thing, I guess, but I didn't know. Probably at least twenty of the guys that I knew, it eventually caught up with them years later and they died from the asbestos.

At the Labs

It was a different life here in town. And then Bell Labs wanted to come in. They wanted to build behind the graveyard. And then the town fathers pooh-poohed it and Bell Labs went to Murray Hill. I understand they bought something like 450 acres up there and they had it all fenced when I went. They opened up the Bell Lab in 1942 and I started there in 1943. Many of the people came in from New York.

My brother Dan worked there. And he said to me, "Why don't you apply?" There were very few jobs out of high school in those days. And all my friends were making about eighteen bucks a week. At the Bell Labs, they paid $25 a week. That was a hundred dollars a month, $1200 a year. That was a lot of money back then, when you could buy a gallon of gas for 15¢.

During the war, we had carpools to get to work. There was no public transportation. If you wanted to go by train, you had to go form Stirling to Summit and then take a bus. As a matter of fact, Italo Quinto used to drive that bus. We went to school together.

I used to like to do things with my hands, like building and things like that. But I didn't have the opportunity to go to college, because there was no money to send me. My father was willing to mortgage the house, but I couldn't see it, so I figured the next best thing is to get a good job, and that's when I went to the Labs. And when I first went in I just got a job as what they call a utility hand. I used to sweep the floors and all that stuff and move the gas tanks around.

I used to be very fascinated when I'd go into the glass shop and watch these old glassblowers working. They were all mostly of German descent. They were scientific glassblowers, but they came from a section of Germany. I used to pay attention and watch them, but they were very tough guys. They wouldn't give you any encouragement. Finally one – I'll never forget it — it was an Irish fellow, he said, "How would you like to learn this trade?" "Well, I'd love to work in here," I said. I suppose there must have been at least eight or ten glassblowers, but they were mostly Germans and they figure what does an Italian know about it? And I used to tell them: "Didn't you ever hear of the Venetian glassblowers?" But this was scientific. So, finally, after a while, they'd sit me down and ask me to wash some tubing, and little by little they took me in. We worked with Homer Hagstrom; he was our immediate supervisor.

Around Christmas, we used to make glass ornaments. I have some at home from the 1950s, when we first started making them. They'd buy special glass and they'd let us blow nothing but Christmas ornaments for a whole week and set them aside. And they had visitors come in on Christmas Eve with the kids. We used to show them how the glass ornaments were made and then we'd finally give them one at the end of the demonstration. Bill used to love that. He'd come in and say what a nice thing that was.

I remember one time they were having a bunch of visitors, from Sony, let's say. When these kinds of visitors came, they liked to have something to present to them. So I made a glass telephone. And I remember when I presented it to the fellow from Japan, I said, "Well, that's very fragile." So, assuming he's going to take it to Japan, I said, "Maybe I ought to carry it over, in case it breaks. I could fix it." He said, "Would you go?" I said, "Hell, yeah, I'd go." So then he said, no, he had an office in New York. They wanted to put it in New York, so that's where it wound up. But I remember we made those. As a matter of fact, I still have one at home. I made an extra one, in case it did break.

We always had the very best equipment, and they used to let us attend symposiums in different parts of the country. All the glassblowers, scientific glassblowers, had our own Society. They called it the American Scientific Glassblowers Society. I remember one of the best glassblowers in my group had left and then wanted to return. I'll never forget, they called me down to personnel. It was a security thing and they wanted to question me about him. And the guy sitting at the desk – I forget what his name was – was a sweeper. I said, "What the hell you doing here?" and he pulled out his identification card. He was with the FBI. He said, "Don't ever tell anybody who I am or what I'm doing here." He was investigating different people. Because they were going to bring this glassblower back, but they wanted to question whatever I knew about the guy. But that really impressed me when this guy, as a sweeper, was one of the FBI security people.

The roofs of the buildings were copper. I often wondered why they had copper roofs on the places. It would be cheaper to put some other type of roofing. I was told, "No," for the simple fact that, with all that electronic equipment, any of the rays would be absorbed [by the copper]. They said otherwise the planes flying over could get screwed up from some of the waves that would come out of the equipment. The partitions were all modular and metal.

Working With Glass

You had to be careful. There were plenty of times you could get a mean cut or a burn. I remember one time I had a great big flask and the fellow said, "Did you rinse it out real good?" I said, "Yeah, it's all done," and I guess he rinsed it out with acetone or something, because the minute it hit the fire, it blew up. And you had to wear your glasses all the time. You could get pretty really well cut up, too, sometimes.

I remember one time we were grinding some apparatus that we had worked on quite a bit and there was a big wheel they called a Tarzan Wheel. We had the abrasive and you would ride it and I think it was a dome with a vacuum. And then the belt slipped or something, and it stopped. So we left the apparatus on top of the thing, because it was stopped, and we put the belt back on, and somebody pushed the button, and it went flying all over. That was the end of about two days' work.

But as a rule, though, it was pretty safe, because that was the one thing that the Labs stressed a lot: safety. And anything that you needed in respect to the safety end of it, they would with no hesitation supply. They checked everything for you. If there were gases that they thought were leaking or whatever, the first thing they would do is send in people in and check it right out. Even if you didn't need it, they'd buy it and make sure you used it, like clothing, gloves for handling hot glass, and all. So that made it pretty safe. Safety was a number one priority.

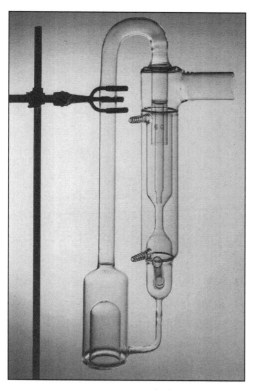

Mercury Diffusion Pump. (Reprinted with permission of David Dorsi)

Say somebody wanted a mercury diffusion pump, which was a very complicated piece of apparatus—with a little luck it would take you two or three days to build one. You'd have to build the sections and anneal it and everything else. Then, other times, we used to go in and put up a complete vacuum station for them. And that would sometimes take a week or more. Some of them were just simple repair jobs that would take minutes. So each job ranged from minutes to weeks. Naturally you took your time and took a lot of pride in what you were doing, but some of the scientists would come in and say, well, they didn't care what it looked like. I said, "I *do* care what it looks like." Inevitably someone else would say, "I don't care whether it works or not. As long as it looks nice, I'm sure it will work good," but some of those scientists, they really didn't care about appearance. They would try to do it themselves and it was in such a hurry. But they would eventually screw up and they'd come back with their tail between their legs and apologize and say, "Well, can you do it for me?" And I would say, "Why didn't you come to begin with? I would have done it faster and better," you know. But they'd figure, well, they'd come out of college, they can do it. It's not that easy.

We worked mainly with Pyrex. There were three or four different Pyrex forms. If they wanted a metal for glass seal, say copper to Pyrex, you had a machinist that would feather the copper edge. Some of them had Kovar metal. And then, later on, they came in with stainless steel and after that they went into quartz. The quartz work was with a higher temperature glass. You needed dark glasses, and you needed good ventilation.

In the beginning it was kind of rough, because your eyes would burn. Later on, they found out we needed special glasses, and even for work with Pyrex they had lenses to cut out the yellow spectrum. And so, when we were working, the color would be easier to work with because it would flow better. It was easier on your eye. But with the quartz they needed a dark glass, almost like welding, because it gave up such a bright light. And then you had to worry about the vapors that come off it. The vapors in the ventilation hood after a while would look like cobwebs. So you had to be careful to avoid silicosis. I mean, you absolutely had to wear a mask and dark glasses.

For Pyrex, the annealing temperatures would be about 550-600°C. But quartz was much higher, a thousand and better centigrade. The working temperatures were different than the melting temperatures. Because we had good ventilation, the heat would travel away from you. Sometimes the heat shield would get in the way and the thing would melt if you were that close. So sometimes the smaller pieces were done by hand, which didn't create that much heat.

They had one department where all they used to do was make kooky things, like hourglasses where the sand would go in reverse.

This fellow and I were playing around with a device that they wanted to use in the tanks for petroleum. They didn't know how much sludge was on the bottom. So we had to come up with a device that had an open tube so that, the minute you got to the sludge from the bottom, it would close off and you could see how much sludge they had in the tank. And that was one of the patents they gave us.

Working With People

Bell Labs had the best people from all over. I mean, they had people come in from all over the world. They'd set them up in a lab: if they succeeded, fine; if not, throw everything out and start over again with somebody else. I remember Sid Millman saying, "If they don't make it interesting for the scientists, in ten years they're not going to have any." And it's true.

We used to have people who would come and ask how long it would take to blow some piece of glass. Well, I said, "How long is a short piece of string?" It's hard to say. Some of those things you can do and it would take a day, other times you could do it in ten minutes, and some of the things that look like they're so simple, you would really struggle at. It was a challenge each time. Some of the researchers would come in with a very fancy drawing and others would just scribble something.

I recall one researcher who worked on the laser, and he would only let the glassblowers in his office. I'll never forget one time he came down from MIT. He used to be at the Labs. And he brought down a laser, and he wouldn't let anybody pack it, so he bought a special ticket on the plane so he could put it on the seat next to him. And when he brought it down, we looked at it and we laughed like hell, because it was a piece of junk. So my colleague took it and threw it in the pail. The guy – I thought he was going to go out of his mind – said, "What are you doing?" I said, "Go have a cup of coffee. We'll make you one that's going to work." In an hour or so we made him a new one. But the only people he would let in his room were the glassblowers. He was a heavy smoker. It looked like a butcher shop, with all the tobacco on the floor. They wouldn't even let the sweepers come in and clean it. I know he got most of the credit for the laser, but there were quite a few people who worked on it.

I remember Sol Buchsbaum. Sol was a little guy, but he was full of hell. I remember one time I built an evacuation station for him, and I don't know what happened. It was a magnetic thing. All the glassware was on the top, and he had the test coil there to test for leaks. And he turns on the magnet, and the magnet was over there and this box was over here, and the magnet pulled that right across and busted the whole glassware. "My God, the look on that poor guy," I thought; so I said, "Don't worry, Sol. We'll do it over again." So we built it again. He said, "How many leaks?" I said, "Guaranteed no leaks," and he said, "Well, what are we betting?" He said, "I'll bet you a good cigar." I said, "Okay." So, seeing he lost, he had to give me a cigar. So he gave me a cigar, and what I did is put his cigar inside a glass tube and sealed it shut. And I gave it back to him, just a reminder that I won. He was a brain that guy. I mean, he moved fast. He went up the line. And I said, "Sol, you're going to be president one day." He said, "Nah." And he got close to it. He was really high up there. I remember he moved fast.

Each scientist was different. We used to come across all of them. It was very interesting. They were a little kooky, a lot of these scientists. They were so smart that they couldn't handle street smarts, but they learned them.

I remember one scientist had rented a car and he was in a hurry to get to the airport and the key didn't fit the car and he got all upset and shook up. So he goes and calls the rental place, they said, "Well, okay, you can break the window and open the car." So he smashes the window — and it wasn't the car that he had rented at all. It was somebody else's car. Another scientist left his car on the ferryboat and took off by foot.

Some were very forgetful. One hadn't been seen for many days and they couldn't find him. And then they finally got a call. He was in Lake Como in Italy, writing his paper. He didn't even tell anybody he had left. He just took off and went. They had to send somebody over, I guess, and pick him up and pay for whatever he did. They used to have to follow some of these scientists. I remember one who they wouldn't even send anyplace because they were afraid that he might be kidnapped and some of his secrets stolen away from him. He was a metallurgist and had something hot going. I don't know what it was. They were an interesting bunch to say the least.

And you had people like Bill Baker. What a person he was. What a mind that guy had. I mean, he would greet you and then right away he'd associate you with hunting or whatever you had in common. And he was now the president and I used to go hunting with him.

We used to do a lot of woodcock hunting and things like that, because we had the dogs and he didn't. He was a good shot, too. He was an outdoorsman. We'd go maybe on a Saturday for a couple hours, and sometimes we'd go during the week for maybe an hour or so. We'd go out at dusk. The woodcock were best hunted just before sunset. They were a nocturnal bird. And he used to like that, because the way they flew and the way the dogs worked on them. It was fun.

The company was very generous in many ways. I remember when the war ended, before they would lay you off, they would give you time off with pay to go look for different jobs. They would seek different places for you to go and interview if you wanted to. Right after the war, I was laid off and went to work for Bakelite in Bound Brook for about two or three years. It was not the same as working for Bell. So then, after the Labs started hiring again, they called me and asked me if I wanted to go back. And I said, "By all means." I grabbed it right away, and I stayed there for about forty years working with the company. I enjoyed it very much, because you came in contact with all kinds of great scientists from all over the world and they contributed a lot. It was great working for that outfit. You really learned a lot and you met a lot of very interesting and very topnotch people.

Retrospect

I left the Labs in 1989, just as Lucent took over. When I left, our glass shop had about eight of us in the chemistry end of it. The physics end had about at least ten or more. But at one time they had maybe fifteen, twenty glassblowers in the whole lab. When they started with the early televisions and all that stuff, they had different groups there. But there were a lot of changes made by the time I left.

I have very, very nice memories of the Labs. And people like Bill Baker were exceptional; you couldn't want to meet nicer people. Some were people who were high up there and they were not pretentious people. Some wanted their work done right though and they were perfectionists. Some of the glassblowers lived in constant fear of making a mistake. "Well, that's what the wastebasket is for. If you screw up, you throw it away." And then they'd start all over again.

EDWIN A. CHANDROSS

Edwin A. Chandross. (Reprinted with permission of Alcatel-Lucent USA Inc.)

From Brooklyn to Murray Hill

I arrived at Bell Labs in the summer of 1959. That was pretty close to the beginning of the big expansion of what had been mostly an applied research lab into a great long-range research institution that was the envy of many companies and universities. What follows is, perforce, a very personal memoir that offers only a single vignette of life in the Bell Labs materials research community. I begin with an account of how I got to Murray Hill. There was an awful amount of luck involved, and some was of the "plain dumb" variety.

I was born in 1934 and grew up in Brooklyn, New York. I am a second generation American whose parents never got past high school. There was great respect in my family for education as the entry to prosperity. My maternal grandmother, who had emigrated from Russia with my grandfather in 1888 as a young woman with at least one child, had learned English well enough to read the *New York Times* on a daily basis. As far back as I can remember, she told me, "Work hard in school, get a scholarship and go to college." Only one of her children had done that; he became a doctor but died of illness long before I was born. All the grandchildren, meanwhile, went to at least a four-year college. That's the wonder of America!

During the early years of grade school, some of us got the idea to go to junior high where we could get into an advanced program where we could make up half a year. That turned out to be a good idea, because it got me on the regular academic track for June graduation. In junior high some of my friends were smart enough to go to the famous Brooklyn

Tech. I knew nothing of such a place and went off to Midwood High School, which is still one of the best in the city. At this time I was interested in both science and engineering. I always liked to build things, so my ambition was to become a civil engineer and design bridges. While I did have a simple chemistry set, there was little that could be done with it, so I was not as exposed to the other sciences.

I took a full academic curriculum in high school that included four years of math and four years of science. I concluded years later that Midwood excelled in its math and English education. Science was relatively weak. Fortunately there was some latitude in science preparation when I got to MIT, but strong English and basic math skills were absolutely needed. Years after graduating, I realized that students in the New York public schools had benefited from the Great Depression of the 1930s. Many of our teachers were very accomplished people who could have had great careers in more lucrative fields had jobs outside of teaching been available.

For my senior year of high school, I had my chemistry and physics requirements left. But, during my junior year, a friend told me of his enjoyment in the first year chemistry class, and that summer I took several books out of the public library. That did it: I was fascinated by elementary inorganic chemistry and drank in great quantities. I started to think about chemistry or chemical engineering as a career. When

I got to the actual class in September, it was clear that I knew more than the teacher, who was pretty gracious in following up when my answers to questions went far beyond what was desired.

Some of us had learned about the Westinghouse Science Talent Search in our senior year and started preparing projects. I was fascinated by reading about the new nickel-cadmium batteries and did a project that compared a simple cell with a conventional lead-acid battery. I managed to get samples of the former from a new company and made the latter from strips of lead. The only support I got from school was a lab bench to work on. Who knew anything about safety glasses or the real dangers of strong acid or alkali? There was essentially no oversight. Several years later I came to understand the importance of surface area and realized that my experiments were quite poorly thought out. However, the first barrier to a prize was an examination that I must have aced. I got an "honorable mention," news of which went to the top 300 universities in the U.S. and generated lots of letters asking me to apply. There was also a New York State chemistry competition and I scored first place in the city.

Of course, there was much thought of college at this time. But there was no place to get any advice, either from family or teachers. In those days one did not ever think of visiting a college to sample it; you simply read all you could. I applied to MIT, Rensselaer, and Brooklyn Poly. It was customary to have at least one that you were sure you could get into. I had a state scholarship that would have paid one quarter of the tuition at any New York school and my parents, who could ill afford it, nevertheless told me to choose the one I wanted. MIT was my choice and gave me a small scholarship. There was some luck in that decision that was not clear at the time.

I arrived at MIT with five of my classmates in September 1951, just shy of my 17th birthday and totally clueless about what my life was to be like. The freshman curriculum, which included Saturday morning classes, was the same for all. I quickly discovered that many of my classmates had been to famous prep schools and had much more advanced classes, including calculus, of which I knew nothing. I was pretty nervous about competing with them, especially after learning of the significant flunk-out rate. It took a whole semester to find out that they were no smarter and did not work any harder and that I could compete with them. Life got less nerve-wracking after that.

When I left high school I was interested in the applications of chemistry to real world problems and thought that I should study chemical engineering. MIT had presentations that introduced students to the various majors. I quickly learned that chemical engineers were called plumbers, and then learned that a lot of their work just involved designing pipelines.

That was the end of my interest in the field, and I majored in chemistry instead. I must point out that chemical engineering changed immensely over the next two decades. Some of that will be mentioned later when I get to the establishment of a chemical engineering department at Murray Hill. I spent a good part of my last twenty years at Bell Labs working on applied research and enjoyed it enormously. In 2007, I completed the circle when I was elected to the National Academy of Engineering based on my contributions to the optical fiber area.

My love for chemical science continued to grow at MIT and I took most of the graduate-level courses offered. I knew that I would have to go to graduate school to get a doctorate in order to have a high-level career, but as usual there was no real guidance available. During my junior year I was told that the research director of American Chicle Company was coming to interview students for a scholarship and a summer job. I spoke with him but never heard anything further. When I went home in May, I had no summer job lined up and wanted one desperately so that I would not have to work part time during the school year as I had been doing. Having nothing else in the offing, I called American Chicle, which was in Long Island City and convenient to my home, to ask if they could give me a job even if I hadn't won the scholarship. I spoke to the head of personnel and she told me that she thought that I had actually won the scholarship and would call me back. That was an agonizing wait but it turned out well: full tuition – all of $900 (that shows just how long ago this was). I also got a summer job working in their factory at $50 a week. They wanted to hire me as a junior executive (whatever that might have been) after graduation, but I had always said that I intended to go to graduate school. Delivering supplies to gum wrapping machines left me with little interest in gum chewing, but I enjoyed the people I worked with.

As a senior I had a hard time choosing between organic and physical chemistry. The beauty of organic structures and the interesting reaction mechanisms that were beginning to be understandable helped me choose. I did my senior thesis on an obscure problem. During that senior year, a friend and I came to the conclusion that we liked MIT and wanted to stay for a PhD. One day I was summoned to the office of the executive officer of the chemistry department. I could not imagine what sin I had committed to justify this but he quickly explained his purpose: "You can go to any graduate school you want, but not here" was the succinct message. At the time I was devastated. It took me some years to realize the wisdom behind that philosophy of not accepting their own undergraduates, but now I am grateful for it.

So, I had to find another choice. My undergraduate thesis advisor, a graduate of Illinois, wanted me to go there. For a native New Yorker, the idea of going to the middle of nowhere was unthinkable. I liked Cambridge and wanted to stay. Fortunately, my advisor's officemate was a Harvard graduate and suggested that I would like it there. He is still a friend and I did eventually thank him for his advice, something that is not common enough. The MIT faculty arranged my admission and full fellowships.

Harvard was very much a sink-or-swim place, with very demanding faculty; they were after all at the top of the heap. The discovery by some of my classmates, all of whom were outstanding undergraduates, that they were no longer king of the hill was a real problem; many of them did not survive the first year. Getting into a research group was not easy unless you went to a junior faculty member, which I did. Some took three years to get a degree, but not many. We typically worked about 80 hours a week, mostly doing experiments. Since we had to pay for one third of what we spent, we were very frugal and that meant a lot of wasted time simply to avoid buying chemicals you could make from cheaper stuff.

I managed to get a National Institutes of Health fellowship after my first year and thus did not have to teach to support myself. In my last year I wanted a car and became a teaching fellow to earn the money for it. I taught undergraduate organic lab, populated largely by pre-meds. Organic chemistry was the killer course for them. If they could not pass it, their chances of medical school were gone. Some of them were pretty grubby in arguing over grades. The takeaway lesson was that it is a good idea to be an educated consumer when dealing with the medical world.

By the beginning of my fourth year I had had enough, feeling that I was not going to learn much more there. Like all Harvard chemists, I knew that the only respectable place for a scientist was in academia and I sought a postdoctoral appointment with a professor at Yale, who was the world's top rated researcher in physical-organic chemistry, the field I wanted to pursue. However, I felt it wise to at least take a look at the industrial world, although I had no interest in the oil or large chemical companies that were sending interviewers to the department. Since I had a long-standing interest in photography, I did sign up to talk to the Polaroid guy. That also fit with my fondness for the Boston area.

I did not go to see the Bell Labs recruiter, feeling that they would have no interest in an organic chemist. I felt foolish when one of my friends, a physical chemist, told me of his conversation with Bill Slichter. He said that they had no positions for physical chemists that year but were looking for organic chemists. I found out that Bill, a Harvard alumnus, was coming back to Cambridge, and I asked to meet him. To my surprise, he turned up in my lab one day and, perched on a stool, interviewed me.

I made a good impression on the Polaroid folks in a one-day interview, and they made me an attractive offer. Meanwhile, as a follow up to my interview with Bill, Bell Labs invited me down, and I visited in December 1958. I had been asked to give a talk on my thesis, and did so on a blackboard. No PowerPoint, and not even any slides or overheads for graduate students in those days. I am convinced that my answers to two questions during that interview got me a job offer. The first was one that posed no problem, even though I had not anticipated it: what would you like to work on? I mentioned chemiluminescence and organic semiconductors. The first was something that had fascinated me since I saw a luminol demonstration during my freshman year at MIT. I wanted to know why a few chemical reactions gave light when most simply produced heat. I had spent many hours bootlegging experiments and reading in an attempt to understand this. I also had some half-baked ideas about organic salts that might carry a current, but I knew essentially nothing about the area.

The second question came from Linc Hawkins, a distinguished polymer chemist, but it seemed to come out of the blue. He asked: what is the benefit of your thesis? Being young (I had just turned twenty four), naïve and honest, I said that it would get me a PhD, and that I would never touch anything like it again. By then I had come to realize that my advisor tended to assign rather unimportant research to graduate students.

Although I had met F. H. (Stretch) Winslow, it was Bill Slichter who called to offer me a job. It was an amazing offer: I could set up my own research program. I had to call him back several times to hear this and believe that it was true. Once convinced, I could not give up the opportunity. It looked a lot better than two years of postdoctoral work followed by struggle as an assistant professor. I accepted the offer, even though the Polaroid people tried hard to sway me.

Now I had to convince my advisor that I had done enough to have a respectable thesis. That took some work but I managed to deliver it in June 1959, too late to get a degree that year. I was a bit nervous about the oral exam when I found out that the two professors who were to question me were known as the toughest examiners in the department (one went on to win a Nobel Prize), but all went well.

I turned up at Bell Labs in August 1959 and joined the polymer research department under Stretch Winslow, one of the nicest and most upright people I have ever known. He was my boss until I succeeded him in 1980, and you could not have wished for better. The new world was quite different from the one I had just left. At Harvard, "polymer" was equivalent to a four-letter word not used in polite company. So much for the narrow-minded, arrogant academic world of the 1950s. Fortunately, this has changed much over the years, and most of my science friends today are in fact in academia.

Bell Labs had pioneered the use of polyethylene in telecom cables, although it had been first used in England. Remember that paper insulation was still used in the 1950s. One of the big problems was the rapid oxidation of the polymer, especially when used outdoors where sunlight accelerated it and led to the polymer just falling apart. Winslow and Hawkins had made great advances in solving this problem. They found that carbon black, the same stuff used in tires, protected polyethylene against both thermal- and photo-degradation. Further, they discovered synergistic combinations of low molecular weight compounds that were the best in the world for protecting polyethylene. AT&T made a fortune from licensing these materials, which were said to be the most valuable patents ever made available. Most patents were covered by the famous 1956 consent decree with the Justice Department that was the price for permission to remain a monopoly.

The department was part of the chemical research laboratory headed by Stan Morgan whose background was in dielectrics, an obvious interest for a telecom company. Stan had decided that he should expand the group's expertise in organic chemistry, as it underpinned the important work of the period, and that is what led to hiring five more people to add to Ed Wasserman, the only chemist with that background in the area. They had intended to acquire only three new chemists, but were too successful with bids. All of us were to start our own programs. We arrived within a few months of each other and quickly realized that we came from a world that no one at Bell Labs understood. This was the beginning of a big expansion of chemistry at the Labs. Besides our organic cohort, there were some excellent physical and inorganic chemists brought in as well. Our department focused on cable and wire insulation. Polymers were quickly replacing older stuff like lead sheathing that had been the standard for outdoor installations. There was also continuing work on polymer stabilization and the ubiquitous issue of cost reduction.

Chemiluminescence: The Lightstick

Our group turned out to be a very interactive collective and we traded ideas and insights freely. Jerry Smolinsky and Tony Trozzolo are still good friends and we keep in touch; Jerry ended up at Sematech and Tony became a chaired professor at Notre Dame. We realized that, as organic chemists at Bell Labs, we had to work on getting outside recognition for our existence and work, and we set out to do that. It took a few years; we got used to being asked what an organic chemist was doing at Bell. I started to look into chemiluminescence and spent a lot of frustrating time spinning my wheels. In retrospect, I could have used some mentoring. But after a few years I made what turned out to be an important discovery: the chemistry of the Lightstick. It was a serendipitous discovery and we did not appreciate how important it would become. Unknown to me, the U.S. Defense Department was supporting a lot of research in the area, aimed partly at finding downed aviators. My family still gives me flak over the refusal of the patent department to file on this. American Cyanamid followed my publication with a good deal of work and came out with a commercial product that is now manufactured by several companies and is well known all over the world. The military is still a big customer. And, of course, many kids carry them in their trick-or-treat quest on Halloween.

This work got me interested in fluorescence and then photochemistry, which remained a major interest for many years. I was well-recognized in that community and there was an important process for AT&T that later came out of this knowledge. This is how investments in long-range research can pay off in ways never imagined. We were considered to be a window on the outside world of science and also consultants. Unfortunately, this philosophy has essentially disappeared from the industrial world.

The question of how the Lightstick reaction produces light is not yet fully understood, some forty five years later. The basic chemistry is the reaction of an oxalate ester with hydrogen peroxide. It forms an energy-rich intermediate, whose decomposition can "pump" a fluorescent species present in the solution to its excited state, from which normal fluorescence is emitted. Electron transfer is probably involved. The beauty of the process is that any fluorescent species can be excited, which is why these devices come in so many colors. I wish I had been smart enough to realize the applications for this chemistry; too bad that the management, much more sophisticated than a twenty nine year-old researcher, did not see them either.

My research in chemiluminescence continued in a different direction. As I mentioned, I was fascinated by the question of why some chemical reactions (mild ones that is, not the violent variety) emitted light. I finally ended up thinking of a general route to such processes. If a chemical reaction occurred faster than chemical bonds could change, there was a chance that an electronically excited state would result. The obvious choice was an electron transfer process that could happen even before two molecules collided. I decided to work with both positive and negative ions derived from aromatic hydrocarbons. This series of compounds begins with two well-known ones: benzene (a toxic liquid) and naphthalene (moth balls). I knew that anthracene, the next larger molecule in the series, was a good choice, so I used a close relative. Adding an electron to form the negative ion was a well-known process. However, the process of forming a positive ion was not well-known. My scheme for making it produced colored solutions that were obviously not stable. However, mixing the solutions of positive and negative ions *did* result in light. So I had to find a better way. I knew that these ions could be formed electrochemically in solution. It was a simple experiment to do this but it never worked well with direct current in the usual manner; positive ions formed at one electrode were too unstable to get efficiently to the negative ones being formed at an adjacent electrode. Now, it is rare that someone will admit to an experiment being done out of desperation, but that is what happened. After concluding that DC electrolysis would not work, I connected the electrodes to a variable transformer and, using low voltage alternating current to drive the two platinum electrodes immersed in a solution of diphenylanthracene with inert electrolyte, lit them up beautifully.

Now, upper management took notice of the ability to get blue light with only a few volts, and did something smart. It was suggested that I meet a newly arrived young electrochemist. We hit it off magnificently and worked together for several years unraveling the basic science. It turned out that DC electrolysis worked fine once very pure materials were used in an inert atmosphere. We realized that this chemistry had promise for a usable display, but that could not be accomplished because the intermediate ions were not stable enough. However, the same basic phenomenon, essentially electron-hole recombination, is now well-established in the solid state and Sony has started to sell an organic light-emitting diode (OLED) television set. That technology will come up later. Bob Visco and I published a number of papers in the area and achieved very good recognition. Bob later went on to become a manager at the Western Electric lab in Princeton and was instrumental in designing the semiconductor plant in Orlando. Sadly, he died of illness when only about fifty years old.

In sum, I did a lot of work with electronically excited organic species and was well-recognized in the outside world for this. But sometime in the early 1980s I found that my interest in fundamental chemistry was lagging. I just could not get excited enough to spend all my time on this. There were a lot of interesting practical problems that got my attention, but which ones to work on and how to pursue them was not clear. Luck, in the form of changes at Bell Labs, would once again enter my story.

Optical Memory

Bill Baker was impressed by the pioneering work in biophysics carried out by a few of the physicists, so, around 1965, he supported the formation of a new department devoted to it. They invited Angelo Lamola, a new assistant professor from Notre Dame, to visit and talk about his work on photo-damage in DNA. He had recently finished his degree with George

Hammond at Caltech in the top group in what later proved to be a renaissance of organic photochemistry. We were all much impressed with Angelo, and managed to persuade him to move to Murray Hill. I had been studying the photochemistry of anthracene, a coal tar hydrocarbon mentioned earlier, and found a simple way (exposure to short wavelength UV light) to convert the known photodimer back to a pair of monomers trapped next to each other in a rigid matrix. I was doing this to try to understand some strange luminescence we had found in the electrochemistry experiments. Around 1968, Angelo and I cooked up the idea of using this reversible process to make an optical memory. He enlisted Jack Tomlinson, an optical physicist at Holmdel, who thought up a completely new way to use this form of photochromism for a memory. Others had studied single-molecule systems that changed color on exposure to UV light and reverted to their original color on exposure to visible light. They are now used in plastic lenses for sunglasses. The problem with those systems was that they were not really all that reversible and reading the image changed it. Jack proposed that we read at a longer wavelength, where no photochemistry could occur, and work on the change in refractive index to make holograms.

We started out thinking that we could dissolve the photodimer in a liquid that could be solidified to a polymer in the Plexiglas family, known to be very transparent and stable. Then, exposure to deep ultraviolet light would create sandwich pairs of monomers, which had a very different absorption spectrum and thus a higher refractive index. An interfering pair of coherent longer wavelength laser beams could then convert some of the pairs back to dimers and thus write a grating pattern that we could detect with red light. That actually worked, but the writing process was awfully inefficient. We realized that we needed a more rigid environment and quickly changed to working with crystals. They worked a lot better but the stress of small molecular motions eventually caused cracking. So, we filed a patent, wrote a paper, and moved on.

I had a great relationship with the guys at Holmdel and our discussions inspired us to move into a new area. It was obvious that the coming era of optical communications could use planar waveguide circuits, which were not known at the time. We came up with some good ideas for doing this in thin polymer films and were quite successful in demonstrating several basic phenomena. This work stimulated many others at labs all over the world. Most important for me was that it was a real introduction to long-range applied research. It was a wonderful collaboration: I did the chemistry, and some basic optical experiments, and Jack Tomlinson and Heinz Weber did the hard physics ones. Jack ended up at Bellcore and Heinz took a professorship at Bern in his homeland. We still keep up with each other. It was fun working like this and we were quite productive.

In the late 1970s, Dave McCall, who was the head of the chemistry laboratory that was focused on the organic materials work – both fundamental research and engineering (there was a lot of very good work on the applied side that was well connected to AT&T manufacturing) – decided that AT&T needed a department devoted to chemical engineering. After all, much of the company's manufacturing was really a large chemical engineering operation, especially the semiconductor manufacturing plants. He got support for this and, since I was already recruiting for chemists at MIT, I introduced myself to their chemical engineering department. I got a chilly reception from the head, who informed me that MIT graduates went into the chemical and petroleum industries where they rose to important positions; they would have no interest in Bell Labs. He turned out to be dead wrong. I was able to bring in some great candidates who went on to very successful careers. Some of the department's alumni are now in academia in very prestigious positions.

Optical Fiber

Some wonderful things happen almost by chance. Pasteur is quoted as saying that "chance favors the prepared mind," and I have no doubt that he was absolutely right. In the late 1970s, Bell Labs was well into a program to develop optical fiber communications. To an outsider it looked foolish to try to compete with Corning Glass, which had a long history in glass and abundant technical expertise. But John MacChesney, a very creative materials scientist, came up with a great way to make the optical fiber perform which became the basis for all fiber technology. It is a rod, about 35 mm diameter and over a meter long. There are now other routes to preforms, but the basic fiber drawing technology remains the same. The preform has an index of refraction that is graded with a parabolic profile, starting with the higher region at the center. Drawing this down yields a thin (about 100 micron) fused silica fiber. But the preforms are expensive and it turns out that most of the glass is there for mechanical strength. So it is overclad with high-purity, fused silica tube that is shrink-wrapped on before drawing.

Obviously, the efficient transmission of light is the *sine qua non* of fiber optic technology. Loss comes from several sources, primarily scattering from irregularities in the index profile and absorption due to impurities. The latter include transition metal ions that absorb light in the near infrared. At that time transmission was done with light of wavelength 1.3 microns. Even a trace of impurity, undetectable by most means, can be disastrous when transmission over hundreds of kilometers is involved.

One day in 1978, I was stopped in the hall by Mark Melliar-Smith, a physical chemist who was a supervisor in the fiber department. He ended up in the semiconductor area and later became CEO of Sematech before going into venture capital and then a lithography startup. Mark asked if we had a spinning band column. As usual, when asked for something, I inquired as to why it was needed. He explained the problem. As I knew, the fiber preform depended on laying down layers of extremely pure silica, which is an inherently low loss material. It is made by "burning" silicon tetrachloride in pure oxygen (this is normally not considered flammable but at temperatures over 1000C it does react to form silica and chlorine). There was a trace of trichlorosilane ($HSiCl_3$) in the commercial material, which was a byproduct of silicon wafer technology. When heated in oxygen, it formed some water, which was eventually chemically bound in the silica. It turns out that the SiOH groups thus formed had some absorption in the infrared at the 1.3-micron transmission wavelength and this loss could easily be measured. They were going nuts trying to figure out how to remove the impurity to reduce the loss. Distillation is not usually effective for impurity levels much under one percent, especially when the boiling points of the two materials differ by only thirty five degrees. So Mark thought of trying the best distillation column he knew of.

I quickly replied that this would not work. The column was basically gold-plated steel and would end up with terrible corrosion caused by the chlorides. I had an inspiration on the spot and said that I knew how to do this easily. This came out of my photochemistry background, which had hitherto been almost entirely directed to fundamental chemistry. All we needed to do was to add a little chlorine to the liquid and shine a light on it. That would initiate the reaction: $HSiCl_3 + Cl_2 \rightarrow SiCl_4 + HCl$. Nothing else could happen and the HCl byproduct is very volatile and easy to remove with a stream of oxygen. Both chlorine and oxygen were already used in preform manufacture, so they were readily available.

Mark introduced me to Bob Barnes, a very talented jack of all trades, who was already working on this problem and in a couple of days we had set up a simple experiment. The light source was a low-power tungsten-halogen lamp from a slide projector. To make a long story short, it worked like a charm the first time out. This is a chain reaction that needs only a few initiating events. Bob used long path length infrared spectroscopy to analyze for traces of the trichloride and found that there was absolutely none left after a short time. The process was sent off to the Atlanta factory and they were using it for production material within a couple of weeks. That got me started on a twenty-plus year involvement in fiber. The chemical engineers took our batch process and built a system that turned it into a continuous one; I think that is still in use for some fiber, although preform manufacturing has changed.

In 1980, Stretch decided to step down as department head and I was promoted. Filling his shoes was not easy. Although I was shifting my interests to more applied work, I did not do much to change the fundamental nature of the dozen or so department members. We did some hiring and eventually moved more into materials, despite some resistance from very academically minded chemists. The materials chemistry world was changing rapidly at that time. Bell had a fantastic reputation in the arena of crystal growth, having invented many that were important in the communications world. Materials science departments in both universities and industry had been focused on metallurgy and ceramics; in fact most such departments with that name had been called "metallurgy departments" until the 1970s.

The big event of the era was the breakup of AT&T in 1984, another day that will live in infamy as far as Bell Labs veterans are concerned. Bellcore was set up to support the operating companies and a large number of materials people were transferred to the new lab, which for a while was still located in the same place. We were shielded from any serious financial impact, as far as I could tell. Budgets were still quite adequate for our needs.

Time to backtrack a bit to introduce the next research area. In the mid-seventies, I learned of a new opportunity in the lithography aspect of semiconductor manufacture. Chips were driving to smaller features, a journey that has continued at a rapid rate. At the time, wavelength of light used to pattern the critical photoresist layer that defines the smallest transistor structure was thought to be a limiting factor. The exposure tools used high-pressure mercury arc lamps and the 365-nm line was standard. Some of the guys I knew in the development area were interested in using shorter wavelength light but there were no resists available. The standard materials absorbed too strongly to allow the exposure light to penetrate to the bottom of the one micron thick resist. At first they wanted to do a simple experiment to show that diffraction would be reduced with a shorter wavelength exposure, but they had no photoresist that was suitable. There had been great progress made by some of my colleagues in electron-beam resists that were used for making the high-resolution masks for exposure. I thought that PBS, an electron beam resist that was transparent at wavelengths above 200 nm, could be exposed at 185 nm, the wavelength responsible for ozone formation. That wavelength could be achieved from readily available low-pressure mercury lamps which were used for that application (those lamps also produce a stronger 254-nm light output that kills bacteria). We set up a near-contact printing test and were pleased that it worked beautifully.

I had some ideas on the design of new resists based on methacrylate polymers that were inherently transparent at 248 nm, a wavelength that was becoming available with new laser sources. In 1980, I enlisted Elsa Reichmanis, a new chemist in my group, and Cletus Wilkins, a new staff member with an organic background in another department. Elsa went on to become a Bell Labs Fellow and department head; she is now a professor at Georgia Tech. Cletus is a professor in Texas. We had a lot of success in the general area and went further in developing analogs of "dissolution inhibitor" resists. There the critical component is a molecule that is very insoluble in alkali but switches to soluble after chemical change caused by exposure. We used cholic acid esters; that is chemistry that I remembered from my undergraduate days and it is natural products chemistry. The human body uses it to solubilize fats during digestion. Whoever would have thought that information from that field would be useful in high technology? This was a good idea, picked up by other workers, but soon supplanted by a very cleverly designed new system that IBM invented. They struck it big and we were jealous.

We were gradually moving more towards research that had at least some relevance to other interests at Bell Labs. In the late 1980s, I encouraged some work in organic materials for nonlinear optics to support the group at the Western Engineering Research Center in Princeton; they were interested in high-speed optical modulators based on materials with low dielectric constants. That looked good at the start but we moved away after a couple of years because it was not attractive enough to pursue.

I continued my interest in optical fiber technology in various ways. Some of my attention was directed to the critical polymer coatings. But the "next big thing" came from John MacChesney, who came up with a new idea for reducing the cost of manufacture. I mentioned that the expensive preform was overclad with a very pure fused silica tube that provided the bulk of the fiber mass. Those were purchased from an outside company that was essentially a sole supplier, and they were very expensive. John thought that he could extend a known idea: starting with a suspension of colloidal silica in alkaline water, it was possible to cause it to gel by simply neutralizing the alkali, which gave you a semi-solid mass. That could, in principle, be dried and then fired to form clear glass. This is, of course, a serious simplification of what had to be done as it overlooks the purification steps needed. But the first big problem was getting a large casting, a couple of inches in diameter and over a yard long around a steel rod that had to be removed. They broke far too easily.

I have done some cement repair at home and knew that adding a polymer to the mix was claimed to give a superior result for thin layers. So I suggested that we try that. The first attempts, all purely seat-of-the-pants choices, failed but we found one that worked and a lot of effort went into developing a technology. I was mostly a kibitzer after the patent filing, although that took a lot of work in itself, and I was at the center. I also had to come up with a believable explanation in the absence of real scientific data. This was turned into a manufacturable technology and work continues on it at OFS Optics, the company that bought Lucent's fiber operations several years ago. The immediate benefit was a very large drop in the price of the cladding tubes we were buying.

Organic Materials

Somewhere around 1992, we became aware of interesting developments in organic materials for active components in electronics. During my interview in 1958, I had mentioned organic semiconductors, but the mantra at Bell Labs had always been that "organics are good for passive components but never for active ones." There was a report from France that showed respectable mobilities for a carefully vapor deposited film of a moderate size organic material and transistors made from it after deposition of metal source and drain contacts. I teamed with Andy Lovinger, who headed a polymer department, and we met with Bob Laudise (he headed the physical and inorganic chemical research lab at the time and had made great contributions over the years starting with the development of hydrothermally growing perfect quartz crystals for oscillators in the 1950s), and we decided that we needed to know more about this. There was also interest in the physics group, so we got an effort started.

Organic Field Effect Transistors (FETs) have continued to attract a great deal of interest all over the world. I have been general chair of an annual workshop sponsored by the American Chemical Society, the Materials Research Society and IEEE for the last few years. There is much work in academia and also a good deal in industry, mostly in Europe and Asia. The most attractive role is in backplanes for large area electronics and in flexible displays, such as for the e-books and fancy cell phones that are just becoming available. Other uses are more problematic, but the field is subject to a hype that pervades everything now.

We did a lot of highly regarded work in this area, and filed some important patents. Another application of organics is in light emitting diodes (LEDs). Several Bell Labs chemists and physicists made great contributions to this area. It has been commercialized in the Sony TV set. At $2500 for an 11-inch screen this seems crazy, but they are selling every one they make. Bigger displays have been demonstrated in hero experiments, and commercial ones are promised. Our entry coincided with a new display effort at Bell Labs with the inflated name "Project Looking Glass." AT&T was already using liquid crystal displays in many products and research in the general area had been carried out in the 1970s. Now the company had been sold on the idea of coming up with its own display technology and a department was formed around 1990. It was headed by a friend who asked me to provide chemical support. That got me into an interesting world that continues to fascinate me. The new department did amazing things, but, after a couple of years, management realized that catching up with the much larger outside effort was not likely to be feasible and the work was stopped.

We had an interesting collaboration with E-ink, an MIT-based startup that had developed an electrophoretic ink for displays. They were interested in using it for flexible displays and needed a flexible backplane to drive them. The organic FETs had poor mobilities and usually required over 50 volts. Fortunately that corresponds to the driving voltage for the display. It uses no power once switched and is ideal for a portable device. We used our expertise in organic transistors to make a good demonstration device. This was driven strongly by John Rogers, an exceptionally creative young chemist. He is now running a huge group in materials at the University of Illinois. A couple of companies have already announced forthcoming products based on this technology. Both the Sony e-book and the Amazon Kindle use this kind of ink and the current backplane is amorphous silicon on glass or metal.

Further Changes

Somewhere around the middle 1990s, the management decided to rearrange all of chemistry and consolidated it. It was never clear that much was accomplished by all of this, but that is the nature of reorganizations. I ended up as a free-ranging consultant and expanded my already wide range of interactions within the company. I had some outside consulting offers but that was clearly forbidden even for areas that had no connection to anything at Bell Labs.

AT&T underwent "trivestiture" in 1996 and Lucent Technologies was established as a separate manufacturing company with Bell Labs as a key part. This story is well-known and will not be discussed here. The Labs continued to have good support until the telecom bubble burst at the turn of the century and then the steady decline of Bell Labs began in earnest. When the "5 + 5" early retirement offer was made to all staff in June of 2001, I thought that it was a no-brainer. They offered to add years to your age and your pension service credit. I was 66 and had no need of the former for getting a pension, but the latter looked like a good thing to have, as I would not bet on being able to last on the payroll for another five years, given what was happening.

I went to see Cherry Murray, the head of physical sciences, to tell her what I was thinking. Having thought about being a consultant (I had been an internal one for years), I figured I could do okay in the outside world. I was surprised to hear that, after the compulsory six-month waiting period for retirees, she wanted to hire me back part-time. I agreed on the spot and that went through. They actually paid me a reasonable rate for the next few years but it stopped when the money really got tight. Meanwhile, I did acquire some major companies as clients and enjoyed that thoroughly. Now that everyone has been hard up for money, business has slowed. In 2008, the ACS hired me as an editor for *Chemistry of Materials*, a very successful journal, and it keeps me quite busy. I continue to be active in the outside world, serving on other editorial boards, such as that for *MRS Journal of Materials Research*, and on advisory boards at universities (MIT and UCLA at present). As I noted earlier, the nicest thing that has happened in recent years has been a few awards: the ACS National Award in Industrial Research and membership in the NAE. I continue to have an office at Bell Labs, but the organization is fast losing its physical science effort and I don't think that will last long enough for me to make 50 years in August 2009. For the sheer enjoyment, I intend to keep professionally active as long as I am able; I enjoy it thoroughly.

To Sum It All Up

There are a few comments that I want to make as I look back on what I have written here. First is to quote my friend Angelo Lamola, who was asked by a colleague at the company he joined after leaving Bell Labs to say what the best part had been. His answer: the lunch table. I think I concur. Those tables often were stuffed with many more people than they were designed for and the conversations emphasized science and technology. The intellectual level was usually incredibly high and you had to watch what you said as you could get corrected quickly. It was a really great opportunity for crosstalk that often led to new insights and inventions.

I have not mentioned the results of the pressure for us to publish and impress the outside world. For a long time there was essentially a friendly competition with labs like IBM: we both wanted the respect of our competitors. Through my technical publications I got to know, and develop friendships with, people in Europe and Japan and was able to travel to meetings in those places. Such voyages were not on my horizon in 1959, when foreign travel was uncommon. At first such travel was hard to sell and I once asked if I could go on my own money. That changed the answer. But as time went on foreign travel got a lot easier for me. Some of the Europeans are still friends and I like to see them when my wife and I go to Europe.

For a long time we had the benefit of visits by world famous scientists, who were happy to come and speak about their work. The research colloquia in the auditorium addressed many areas that were not covered locally and they contributed much to the intellectual environment. When I visit a university now and am exposed to such things, I realize again the joy of being in a community of accomplished and intellectually curious people.

I am very unhappy that there is so much that I cannot write about in this limited essay: all the marvelous work associated with the materials world at Bell that overwhelms what I have described here. There were so many bright, creative people who made it a pleasure to come to "work" and a challenge to keep up with the competition. Over the years I had the pleasure of collaborating with a great many people, only a few of whom are mentioned here. They taught me a lot about many areas that were absent from my education, both in science and engineering, and they enriched my life and my abilities enormously. I wish there were some way to acknowledge them all.

To sum it all up, I ended up at Bell Labs without really knowing anything about what I was getting into. It was, to use a worn cliché, "one hell of a ride," one that matched my interests well. I have not included the various positions offered to me elsewhere along the way and saying "no" was a good move for every one of them. Looking back on it all, I often comment that we did not know how good we had it. Industrial research of the sort at Bell Labs has largely disappeared. It has been pointed out that this was affordable only by companies that were essentially monopolies.

Nonetheless, the United States is the poorer for it. I say that not just because of the work done at labs such as Bell, but because of all the talented scientists and engineers who got their start there and went on to do marvelous things elsewhere. Universities and Silicon Valley are full of Bell Labs alumni, but the well is dry. *Sic transit.*

ITALO QUINTO

Italo Quinto. (Reprinted with permission of Italo Quinto)

Note: Italo Quinto was interviewed by Richard Q. Hofacker on August 16, 2007. A. Michael Noll then edited this interview into prose, retaining the voice of Mr. Quinto as much as possible.

From Italy to the US

I was born in 1923 in Lenola, Italy, which is midway between Rome and Naples on the west coast of Italy. My mother, father and I left Italy on an Italian ship bound for the United States when I was six months old. My father had come to the United States earlier, around 1902, and had become a citizen, but had returned to Italy to find a bride. We landed in New York and we went to live in Plainfield, New York. My brother Ermando was born probably in 1928, and my brother Frank followed in 1930.

My father was an electrician in Italy where he was already earning good money. In the United States, however, he went to work for the railroad; I believe it was the Jersey Central. My mother had died in 1925, and I was put with one family to another until my father remarried. He remarried another Italian woman and brought his second wife here to the United States along with and my one brother that was still in Italy. He started a family all over again. My father had two boys and one girl with his second wife.

When I was about five years old, or perhaps before, we moved to Stirling, NJ where I started school. School was of course conducted in English. I would not learn the Italian language. I figured I'm over here. I used to laugh at my father when he'd try to get me to learn. I did not attend high school and started to work when I was sixteen. I graduated from the eighth grade in 1940 or 1941. I started work in a knitting factory, making women's evening gowns and sweaters. After a year, I was laid off and went to work making tassels for coffins and other things.

Gunner in World War II

My older brother was drafted in 1939 into the Army. He went on the invasion of North Africa on D-Day plus two. I don't know if he saw much action, because he didn't say much. Yes, I think he lucked out on that. He was at Casablanca most of the time. Then from there, after the invasion of Southern Italy, his outfit went into Italy, and they were up in what they call Santa Maria, Italy. That's just above Naples. His company was there until the invasion of southern France, but he wasn't in on that invasion either.

With the war-based economy gearing up, I worked on making piston rings for aircraft at the U.S. Hammer and Piston Ring Company in Stirling. Then, in 1943 at age eighteen — going on nineteen — I was drafted! We went to Fort Dix to be checked out – a physical. About three days later, they needed a shipment to go to the west coast, but the people that were supposed to go to had measles in that barracks, so they got held back. So thirty of us guys went to the west coast instead. We were put in with the infantry, the National Guard 30th Division. That was the Canton, Nebraska, National Guard outfit that was out on the west coast. That's how we started our infantry. Then the whole unit got shipped to Alabama.

I had problems with my toenails and ended in the hospital. They never really healed with all the walking. We wondered why all the walking and didn't realize that they were preparing us for the European war. And then an order came out that anybody who wanted to transfer to the Army Paratroops or the Rangers and the Air Corps could apply for transfer. And a lot of us said, "To hell with this walking." Many transferred into the other branches. This was now July of 1943.

I didn't like the Rangers. As a matter of fact I didn't like flying. When we applied to transfer, they shipped us all out to what we asked for. I went to Harlingen, Texas, the place for gunnery school. I had never been up in a plane. One day I figured I've got to go up some day, I've got to see how it is. I went to the air base, and they had pilots that would practice landing and coming down. And I said to the pilot, "Hey, could I come and take a flight?" "Yes, come on." I ran away, in fact, to the base. So after the third time doing that I convinced myself that I had to fly sooner or later. We were flying in a single-engine trainer.

I trained in gunnery for six weeks, firing out of a turret in a twin-engine plane. The gun fired 700 rounds per minute, and we fired at tow targets. There was a long sleeve, and we'd shoot at that. And to qualify you had to get a certain percentage of them. They had a marker on them. They also sent us to another place for clay shooting with a rifle. We went around in a two and a half ton truck and shot from the truck. The truck was moving, and there was somebody launching the clay pigeons. After six weeks of gunnery training, we went for three months of aircrew training. We went to Florida, outside of Tampa. I think it's an international airport now. They put the crew together as you reported in. The clerk looked at me, and since I was a small guy he assigned me as a gunner in the belly turret of a B-17. That's the way they did it.

After the training we were ordered to England at the end of April 1944. We were shipped up to Savannah, an Air Force base, and when we got up there, we got the plane we were going to take. We did a few flights to make sure everything was all right. But the weather changed, and we got delayed about two weeks.

We flew the B-17 over to England, landing first in Labrador and then Iceland. We went into Scotland – the Prestwick Air Base. Today it's a commercial airport. They took the plane away from us when we got there. When you got to your assigned base, they gave you another plane. We renamed our plane "I'll Get By" from a popular song then, "I'll get by as long as I have you." We went to another base for a week of indoctrination. Our plane already had two or three missions on it, and we finished it up. We took that plane on a total of 114 missions. It was the most missions any plane did in that particular bomb group – the 390th

Bomb Group of the Eighth Air Force. Up until I got there, you did twenty-five missions and you were sent home. After D-Day, the missions got so easy General Doolittle wanted us to fly until we got shot down. The morale went downhill and they had to do something, so they came out with a new order that you have to fly thirty-five missions before you could go home. That made the guys a little bit happier.

We flew our first mission on June 4, 1944, right before D-Day. The target of our first mission was the area where the Germans had all those robot bombers with no pilot, the V-1s. We were bombing up there simply to let the Germans think that was where the invasion was going to be. D-Day was our third mission. We went in and bombed a bridge and roadway in France. There were between a thousand and twelve-hundred bombers spread out all over that day. To bomb the continent, we would circle over Great Britain to gain an altitude of 15,000 feet and then fly a straight line toward the target. At that altitude, we wore oxygen masks. As soon as we dropped the bombs, we would dive 1,000 feet, make a left turn, and get out. When we got back to base, we got debriefed, had a bite to eat, and then went to sleep for a while. It was cold on top, but we had an electric heating suit with a fur-lined jacket and pants over it. Once on top, I spent anywhere from eight to twelve hours in the turret.

In the later part of August 1944, we started flying over Germany as soon as the fighter planes got the wing tanks for extra fuel. Soon as we got a foothold in Germany, we started going out farther and farther, and the missions started getting longer and longer. Our bombardier got killed on a mission from antiaircraft flak at around 20,000 feet while we were still climbing. We got hit right above the nose. When the shrapnel started coming into the plane it came right through the metal and the plastic. The bombardier had on a helmet and a helmet liner, and the flack was that powerful it went through both of them and lodged into his neck. The plane was still able to fly, but we abandoned the mission and went back to base. There were around 1,500 planes in the mission that day over Düsseldorf. We did only day bombing; the Brits did the night bombing – they were the heroes.

I went on what was supposed to be a highly secret mission in August of 1944 – well, not only me, our base. It was to fly from England to Russia, do a mission, go back to Russia, go from Russia to Italy, and then come back to England. We were briefed we were going to land at a highly secret air base in the Ukraine. That was hardly secret. That night the German fighters came over and dropped bombs on us.

In total, I flew in thirty five missions over Germany and received five battle stars and the Distinguished Flying Cross. Then we came back home on a Liberty delivery ship. I think we took thirty days to get home. Our ship landed in Newark. We slept at Camp Kilmer overnight, then they took us by bus down to Fort Dix, where they gave us a furlough. We got a thirty-day furlough. I didn't luck out at that. I got orders to go to Redding Air Force Base. I put in for Langley Field to go to radio school. There I got assigned to the motor pool.

About every three weeks I got to come home. And this is what annoys me about these people saying, "You World War II guys came home to parades." Everybody didn't come home to a parade, and everybody didn't have it easy either. I was at Langley until peace came. I was out in the piney woods of Mississippi when the news came in that the Japanese had surrendered. That sounded good to us, because we didn't know where we would be going except that it would be in the Pacific some-where. We were sure of that, because Germany had surrendered. And then to hear that the Japanese had surrendered meant that they might not need us anyway. I was discharged September 3, 1945. I married in 1947 and had three children. My wife passed away in 1995.

Professional Driver for Bell Labs

I did not go looking for a job right away upon getting home. I went drinking for four weeks. But money was getting short, so I got a bus driver's job for the

Summit-New Providence bus line. We took kids to school and also people who worked at Bell Labs in Murray Hill. A lot of the people coming in from the city would get off at Summit, and they'd get on a bus that was waiting outside the station and ride up to Murray Hill. The fare was a dime.

The Summit-New Providence bus. (Reprinted with permission of Passaic Valley Coaches Inc.)

It was driving that bus that I first met Dr. Baker, who knew who I was. My bosses at the bus company recommended me to Bell Labs, and I started there in March of 1948 at the motor pool. I took a driving test, which consisted of driving a truck, slowing down and stopping, and back shifting. I had some experience driving a truck down at Langley Field.

I drove trucks for two or three years, and then became an executive chauffeur at Bell Labs in 1967.

I met Dr. Baker again when he was president of the Labs. He was great. I drove him and other executives all over, mostly to New York but sometimes to Philadelphia. I once drove Dr. Baker up to Rhode Island. He had to make a speech there. He wouldn't go by plane. He wanted to go by car. He was an incredible person. Dr. Baker could be in a huddle talking to people and he could be in rags, and nobody would think about the rags – all they would think of was that they were in the presence of a great man.

There were six executive chauffeurs in all. Our normal workday was eight hours. We had to make two trips. Some went early in the morning, and then the driver would come back and take whoever else was left to go in the afternoon. My salary at the beginning was $42 a week. My final salary was $62,000 a year. We had one chauffeur who used to like to brag a lot, and so he bragged about how much money he was making. You've got these engineers, guys with higher education, and they're making nowhere near it. And they go to their boss and say, "Guy drives a damn car, he makes X amount of dollars and we don't," and their bosses started getting on our bosses – which is only right – "Why don't your guys shut up?" But that guy, he had to brag.

I remember driving Dr. Baker to St. Peter's College in Jersey City for their 100th anniversary. It had to be 1983 or 1984. Well, when we got to where the banquet was going to be, they had the whole block blocked. Dr. Baker asked, "Well, what are you going to do now that you're dropping me off here?" I said, "Well, I'm going to park down there, and I'll wait for you there." "Oh no, you're not. You come in." I said, "Yeah, okay." I parked the car and stayed in the car. Pretty soon Dr. Baker came out and he demanded that I go in. When I got in another surprise was the only chair that was left was a chair at his table. And he told me, "You sit there." Fortunately, I had on a suit with a dark jacket and pants, and not a uniform. In Europe they were having problems with executives being kidnapped, so Bell Labs stopped having us look like chauffeurs when we drove the executives.

I continued to drive Dr. Baker even after he retired. He was entitled to chauffeur service because he was still representing Bell Labs or the AT&T system. He'd be going to the White House, or he'd be going down to Trenton. I retired from the Labs at the end of 1984, after working there for twenty-two years. Even after I was retired, Dr. Baker always asked this one chauffeur how I was doing. Which was nice, you know, because a lot of them could not care less about you after you are gone. But Dr. Baker was not that way.

MAN MOHAN SONDHI

Man Mohan Sondhi (front) with (l to r) Larry Rabiner, Fred Juang, and Steve Levinson. (Reprinted with permission of Alcatel-Lucent USA Inc.)

Beginnings: India to the United States

I was born in 1933 in a small town called Ferozepore in Punjab, India, just a few miles from what is now the border between India and Pakistan. My childhood was spent in Delhi, where my family settled when I was just a couple of years old. At the time (and through August 14, 1947) India was a British colony. British rule was a mixed blessing. The British left India with some very worthwhile things – e.g., a railway system, a civil service, a military, and the English language. However, they also ruthlessly exploited the natural resources of India and treated Indians as second-class citizens in their own country. By the time I was aware of my surroundings this discrimination no longer existed, at least not overtly. Indeed there were not that many British people left in India by then. But my parents' generation was quite strongly affected by it and my father related many stories of indignities suffered by Indians during his formative years.

One of my memories of early childhood includes evenings – especially during the summer months – playing in the parks surrounding India Gate in New Delhi. The "Gate" is really an arch, sort of like the arch in Washington Square Park in New York. Looking through it from one side, one could see a statue of King George VI of Britain. (That statue no longer exists. I assume it was removed in nationalistic fervor after India's independence.) Another memory is of attending the annual performance of Ram Lila. This is a re-enactment of Ramayana, which is the epic story of the life of the mythical king Rama and his conquests over the demon king Ravana. The drama was presented on floats that circled an audience of thousands of people in a huge park. It took many evenings to show scenes from the book and, on the last day the performance, culminated in the lighting of huge effigies of Ravana and a couple of his brothers. To me, as a child, the three effigies looked like skyscrapers. In reality they were probably twenty or thirty feet high. They were filled with fireworks, mostly loud bombs and firecrackers. I remember being thrilled by all the performances but terrified by the loud noises on the last day. To this day I watch fireworks on the fourth of July only with earplugs to shut out all the loud bangs.

My schooling started at the age of three years, when my parents admitted me into kindergarten at Queen Mary's School. That was basically a girls' school, but it was co-educational in kindergarten. Upon finishing kindergarten I transferred to a co-educational school called Cambridge School. That school prepared students for the Cambridge School Leaving Certificate in eight years. This certificate was administered by Cambridge University of England, and was evaluated to be equivalent to an Indian high school diploma. After high school, for quite some time, if people would ask me what I planned to do next, I would say "I will get a bachelor's degree in commerce followed by a degree in law." This was, I assume, because my father had a law degree (although he never really practiced law) and had been a rather successful businessman throughout his life. However, when my father had a serious conversation with me about my ambitions and asked me in private what I really wanted to do, I expressed a strong interest in science. Fortunately, he supported me. In hindsight, I think I would have made a terrible businessman.

With the Cambridge School Leaving Certificate in hand, I was eligible to enter college in 1947. I joined Hindu College and registered for the B.Sc. Physics (Honors) course of Delhi University, with mathematics and history of science as subsidiary subjects. The physics honors course was a comprehensive three-year course that covered all aspects of physics at an undergraduate level — mathematics applied to physics, electricity and magnetism, optics, thermodynamics, atomic physics, etc. I finished the course in the first division, and was ranked fourth in my class of about 40. I like to think I would have ranked first, were it not for the fact that my lab notebook was messy and not well kept. Unfortunately for me, considerable weight was given to the way records were kept of all the experiments we performed during the three years of the course. I did not do that well at all. Poor record-keeping haunts me to this day!

After my physics degree, I enrolled for the diploma in electrical communication engineering at the Indian Institute of Science in Bangalore. This was (and is to this day) a very prestigious institute that was established by the great industrialist and philanthropist J. N. Tata. (The industries and institutions established by him and later by his family are some of the best in India and are now well-known worldwide.) The diploma of the Institute was well regarded in India, and, although it was not officially a degree, it was considered similar to an MS degree.

After finishing my diploma, I stayed on at the Institute as a research associate for a year. I wanted to continue further studies in engineering, but in those days (mid-1950s) no graduate program in engineering was offered anywhere in India. My research advisor and mentor, Dr. B.S. Ramakrishna, encouraged me to go abroad for further studies. He had recently returned from the U.S. after graduate studies and suggested that I apply to some universities there. So I decided to go to America for graduate studies provided I could arrange financial assistance. The University of Wisconsin at Madison was one of the universities that offered me a research assistantship. So, in 1954, I embarked on the great adventure of a sea voyage bound for New York. (Air travel those days was by propeller planes and only for the very rich.) The first part of the voyage was to London via the Suez Canal and Gibraltar. The second part was from Southampton to New York after a week's stay in London. While I was on the ocean on my way to America, my very first scientific paper, "Vibrations of Indian Musical Drums Regarded as Composite Membranes," written jointly with Dr. Ramakrishna, appeared in the *Journal of the Acoustical Society of America*. I arrived in

New York just about the time that Senator Joseph McCarthy was beginning to fall from grace.

For my first year in Madison, I stayed at the International House, which was an interesting experiment sponsored by the university. It housed thirty university students in fifteen rooms, in such a way that each room had one American student paired with a student from a foreign country. We had all our meals together and the interactions in this international gathering of students were very interesting and friendly. My first year in Madison gave me many experiences for the first time. One of these was the weather. I experienced a snowfall for the first time in my life. And the minus 17 degrees Fahrenheit temperature that winter was quite a drop from the 110-degree temperature of Delhi earlier that summer. Naturally I was also exposed to some cultural changes from what I was used to. I remember being aghast at a student very irreverently talking of his father as "my old man." And I was amazed at the trustworthiness of people in the Christmas season, when they would simply leave a package for the mailman beside a mailbox if the package was too big for the box. In Delhi, the packages would not have lasted there for more than a few minutes before being stolen. (To be fair, I guess they probably would not last in New York, either.)

After the first year I was fortunate in being nominated for the Knapp Foundation fellowship. This, too, was an interesting experimental program. About a dozen graduate students (I forget the exact number now) were selected to live together in a mansion that was once the governor's residence. The selection was such that no two students were from the same academic discipline. Once a month we had a meeting of all the students and their respective major professors, along with a few deans. At each such meeting one of the students gave an hour-long presentation of the subject he was exploring for his thesis. The idea was to give the students an exposure to current research in a wide variety of subjects. It was altogether a very stimulating exchange of ideas. The two years I spent in that program were the most enjoyable of the time I spent at the University of Wisconsin.

The training I had in my undergraduate studies in India turned out to be excellent preparation for my graduate studies at Wisconsin. Indeed, I think I was much better prepared than most of my fellow students from American schools. I had little difficulty with the courses I needed for my graduate degrees, so I had plenty of time for social activities. I became very interested in American folk music and joined the folk music group at the university. I learned to play the mandolin and also learned the Scruggs style of playing the five-string banjo. I remember many activities associated with folk music during those years, but two events particularly come to mind. One was when our folk group arranged to invite the famous folk singer Pete Seeger to give a concert at the students' union. Seeger, like many other entertainers, had been blacklisted by Senator McCarthy's Senate Permanent Subcommittee on Investigations. So his visit became a tumultuous affair, and many students paraded up and down near the union with big posters that read: "We don't want any commies here." The paranoia created by McCarthy was clearly still around, even in the usually liberal setting of a university. I was shocked at the hostile reception of this gentle, lovable folk singer. Fortunately, the demonstrations stayed nonviolent. The concert took place without further incidents and drew a large gathering in spite of the demonstrations. After the concert I had the good fortune of spending a delightful evening with Seeger at the home of friends who were hosting the singer overnight. The other event I vividly remember is a visit by the poet Carl Sandburg who gave a reading of some of his poems in the university auditorium. After his performance, our folk group (including me) entertained him with some songs at a small gathering in the students' union.

By 1957, I had received my MS and Ph.D. degrees and, as an exchange student, was eligible to seek employment for "practical training" for eighteen months. I had a great respect for Bell Laboratories. I had read Claude Shannon's papers on information theory and knew that transistors had been invented at Bell Labs. Two graduate students from our department had recently joined the Labs and spoke very highly of their experience. And, of course, I had heard of several famous scientists who had worked there or were then working there. So Bell Labs was definitely a place I wanted to join and it was one of the labs to which I applied for a job interview. I had very good interviews with several groups there and at least one of them seemed to be interested in offering me a research appointment. However, there was a snag. Bell Labs those days was a defense establishment, and it turned out that they could not employ me even temporarily unless I obtained US citizenship. I got a very nice letter from the employment office of Bell Labs saying that a couple of departments with whom I had interviewed were interested in me. However, before they could offer me employment I had to at least declare intent of becoming a citizen, and the labs would apply on my behalf and complete all the paper work necessary. At that time I had no intention of permanently settling in the US and all along my plan had been to return to India after my practical training. So, reluctantly, I had to decline the offer. Fortunately I had a couple of other offers and selected a company in Wisconsin whose research staff was developing analog computers for use on jet aircraft as pilot aids.

Upon my return to India, I joined what was then called a "scientists' pool" that the Indian government had established as a stop-gap measure to help people returning after studies abroad. The government gave me a small but reasonably adequate stipend to work at an institution of my choice, while I was waiting for a suitable permanent position. I chose my old department at the Institute in Bangalore and taught a couple of undergraduate courses there. Within a few months I accepted an offer to join the Central Electronics Engineering Research Institute, which is a government-run institute in a small town called Pilani, not far from Delhi.

To Bell Labs

Physically speaking, life in Pilani was very comfortable and pleasant indeed. I was able to rent a small house and employ a live-in cook who also did the household chores. And the people at the institute were mostly young and bright and enthusiastic and fun to interact with. From a professional point of view, however, I was not too happy. There was very little that was going on that was of direct interest to me and nothing where I could use my talents. I felt that I had to be in a more stimulating environment. Also, my sister and parents (that is, all my immediate family) were now in the U.S., and planning to stay there permanently. So, less than a couple of years after returning to India, I decided to leave once more. I wanted to return to the US but, if I remember correctly, after being there on a student visa I was not eligible to return for two years. So I applied to the University of Toronto and was offered a faculty position in the electrical engineering department. I taught a couple of graduate and undergraduate courses there for a year. The university wanted to extend my appointment but, before settling in academia, I very much wanted to again explore the possibility of working for some time at Bell Labs. I wrote a letter to their employment office basically asking if their earlier letter had been a polite refusal or if there was genuine interest in offering me a job. The reply said that they really had been interested in me in 1957 and invited me to come for a new round of interviews. After a series of interviews in 1962, I finally joined Bell Labs in July of that year. Of the couple of departments that showed interest in me, I chose the Information Principles Division. The main reason for my choice was that the various research problems being investigated there covered a wide spectrum of topics including two that were especially appealing: speech signal processing and acoustics. I was also partly influenced by a colleague of mine from the institute in India, B. S. Atal, who had joined that division a year earlier.

I must confess that when I joined Bell Labs, I had intended to stay there for just a few years, to gain some experience in industrial research, and then return to academia. The research environment, however, was so attractive that the few years stretched out to thirty-nine. Let me share with you some of the allure and excitement of working at Bell Labs that was responsible for this change in my plans.

Fortunately, a large part of my career at Bell Labs was during the era that has been called "the glory days" of Bell Labs. I am not sure when that era began – presumably just after World War II. However, the end is easier to identify. Slow erosion started in 1984 when AT&T was broken up, even though Bell Labs itself was still kept in one piece. The erosion picked up speed in 1996, when Bell Labs was split up into Lucent Technologies and a new AT&T Research Lab. And at least for the area of Bell Labs where I worked, the end came abruptly in 2001, with what I call "an epidemic of encouraged retirement." But I am getting ahead of my story. Let me tell relate some of the qualities of Bell Labs that made it such a unique place and, wherever possible, give the reader some selected examples from my own personal experience that illustrate these qualities.

An Academic Atmosphere

One of the first things that impressed me at Bell Labs was the essentially academic atmosphere of the place. That degree of academic freedom was something I had not anticipated in an industrial research organization. And, to add to that, the Labs had resources in equipment and support personnel that far exceeded those at a university. Also, top-quality computing facilities were available. (By today's standards, of course, the computing facility at Bell Labs in 1962 is laughable.

All computations were done on an IBM 704 computer in batch mode. The computer had a memory of 32K, and it occupied a large room in the only air-conditioned building on the campus. Programs were punched on Hollerith cards and sent to the computer via mail carts. However, by the standards of the day, the facility was top class.) Another thing that struck me immediately was the completely free atmosphere. Here were world famous luminaries by the dozen, who were not barricaded behind outer offices or their secretaries. You could just walk up to their offices and they all welcomed you to discuss problems with them.

Diversity was another remarkable aspect of the Labs. The department that I joined was ostensibly concerned with various aspects of speech signal processing, coding, transmission, etc. Everyone in the department was interested in one aspect or another of speech. But as far as I could tell, almost none of the members of the department joined the Labs with prior credentials in that area. There were a couple of electrical engineers, a physicist, a physical chemist, a mathematician, and so on. And yet, together they were able to produce some of the best work in the field. This approach is in sharp contrast to the modern idea of hiring people who would "hit the ground running." Yet another aspect of diversity is worth mentioning. It was the acknowledgement that people have various motivations for doing research. Some (like me) approach it as puzzles to be solved;

others to make a fundamental discovery or to make a lasting contribution to society; yet others to invent commercially successful products; and so on. It was recognized that good research can be done with a variety of motivations.

Freedom to choose the line of research one wished to pursue was one of the most enticing aspects of working at Bell Labs. I can say without hesitation that in my thirty-nine years there, I was not once asked, let alone imposed upon, to work on some specific problem. And I am sure this experience was shared by just about everyone in at least the basic research part of Bell Labs. The idea was for managers (e.g., department heads, directors, etc.) to provide a problem-rich environment and then let individuals decide what they wanted to work on. Some very broad directives obviously existed, and I assume straying too far from them might have been discouraged.

Collaboration among people in very different parts of the Labs was encouraged. My collaboration with several members of the mathematics department is an example of this. That collaboration yielded several papers in queuing theory, which was definitely not something of direct interest to the department that employed me. (Incidentally, one of those papers, written in collaboration with David Anick and Debasis Mitra, actually turned out to be cited far more than any of my papers in acoustics or speech.)

Emphasis was on doing good science and creating new knowledge without the immediate goal of practical application. Not all research was aimed at developing products. The thinking was that, with several competent people doing research with a variety of motivations and in an environment rich with relevant problems, some of the research was bound to result in good products. This attitude started changing slowly in 1984, when there began to be more emphasis on product-oriented research. Some revisionist claims have been made to the effect that research at Bell Labs was always directed towards a product, and the invention of the transistor has been cited as a prime example of this. I do not agree with that assessment. And as far as the invention of the transistor is concerned, I distinctly remember a symposium in honor of the 50th anniversary of its invention. Several scientists who had participated in that activity in the 1940s were at this symposium. And several of them made statements to the effect that the study of semiconductors was presented to them as an area worth investigating with the potential of yielding breakthrough products. However, their charter clearly established a program to study semiconductors and develop an understanding of their properties, without the specific aim of developing a product.

Finally, I might mention that there was a fun-loving atmosphere at the Labs cultivated, no doubt, through the policies of W. O. "Bill" Baker and other upper management. Once, when Baker came to our department, he asked a visiting Japanese scientist if he was having fun during his visit at the lab. The scientist took this as a cue to tell Baker how well he was being treated and how interesting his work was, and so on. After listening to this for a few minutes Baker said, "Great! But what I asked you was if you are having fun." Another incident that illustrates this is when someone in our speech recognition group brought in a toy tank that could be controlled by voice commands. Everyone was having fun playing with it in the corridors and at one point the tank rolled into the executive director's office and "fired" a shot. Rather than being annoyed at this, the executive director joined in the fun. Then there was Ron Graham, who was always entertaining with his gadgets and juggling and tricks. Following are but a few examples from my personal experience that illustrate the emphasis on creation of knowledge, and the emphasis on long-term research at Bell Labs.

Speech Science and Echo Research

A problem of interest in speech science is to determine the continuously changing cross-sectional area along the vocal tract in the process of producing speech. I was made aware of this problem early in my career by James Flanagan. Early procedures for tackling it used x-ray movie cameras to record side views of the vocal tract and, from them, estimated its three-dimensional form. This method is undesirable for several reasons, the most important being the need for prolonged exposure to x-rays. Manfred Schroeder suggested using acoustical measurements to determine the shape and derived a valid method when the tract is only slightly perturbed from a uniform cross-section. B. Gopinath and I then got intrigued by this "inverse" problem and were able to derive a complete mathematical solution. Later, in collaboration with Jeff Resnick, and then with Juergen Schroeter, I was able to reduce the method to practice and actually produce movies of the dynamic variations of the vocal tract. This work did find recognition in some areas of the scientific community but ultimately turned out not to be too useful for speech work. Yet its scientific merit was appreciated and the work was supported at Bell Labs for several years.

As an example of emphasis on long-term research, I might mention my work on echo cancellation. In long-distance telephony, echoes are generated at points of impedance mismatch, mainly at the interface between the long-distance network and the local network. Several satisfactory methods of combating these echoes had been invented over the decades since the advent of long-distance telephony. However, in 1962, one week after I joined Bell labs, the era of satellite telephony was ushered in with the launching of the first communications satellite, *Telstar*. It was anticipated that satellite communication would ultimately occur via geostationary satellites and these produce round-trip delays of as much as 600 milliseconds. With such long delays the known methods of dealing with echoes were shown to be quite unsatisfactory. A new method was needed, and John Kelly started thinking about it in the early 1960s. He proposed the device now known as an echo canceler and invited Ben Logan and me to join him in developing it. Anthony Presti and I made the first (analog) prototype of the canceler around 1966. Due to Kelly's untimely death, I ended up writing the first paper laying out the mathematical properties of the device in 1967. In the 1970s, Debasis Mitra and I collaborated in deriving several basic properties of the canceler.

The important thing to note here is that, while all this work was being supported, it was not clear if and when a commercial application of the device would be realized. The analog implementation was prohibitively expensive and integrated circuits were still in the future. Nevertheless, we were encouraged to work on "impractical" ideas under the assumption that eventually digital technology would make them economically feasible. It took over fifteen years for commercial application of echo cancelers to begin, with the implementation, in 1980, of a canceler on a chip by Don Duttweiler and Y.S. Chen. Later, of course, the device was deployed by the millions on the telephone network and yielded more than a billion dollars of revenue to the owners of Bell Labs.

Another example of aiming at a distant future is provided by a more recent application of echo cancellation. During the 1990s, I started thinking about the problem of canceling echoes in a stereophonic teleconference. It turns out that cancellation of echoes in such a multi-channel situation is qualitatively different from cancellation in the single-channel case. In collaboration with Dennis Morgan, Jacob Benesty, and Joseph Hall, I spent a considerable amount of effort on this problem, and we were able to come up with some interesting and useful solutions. Again, this work was encouraged even though no immediate use was expected. More than ten years later, there are still no commercial applications of this work. To date it remains a conjecture, although I believe a good one, that eventually most teleconferencing will use stereophonic (in general, multi-channel) transmission and will therefore require stereo (or multi-channel) echo cancelers. I could give several other examples from my personal work in other areas such as speech recognition and speech synthesis, but the above examples suffice to illustrate my view of Bell Labs' commitment to long-term research aimed primarily at creation of knowledge.

The Future of Industrial Research

It may be argued that this attitude towards industrial research was possible only because of the special, regulated monopoly structure of the old AT&T. That is probably true. However, it was clearly due, in large part, to enlightened leadership of the type provided by Bill Baker. It may also be argued that in order to show an annual profit, business value becomes of paramount importance, and, as such, industries cannot nowadays afford the luxury of such research. There might be some truth in that assertion too. However, I can cite several ideas that came out of the work during the "golden years" whose commercial exploitation (forbidden under the regulated monopoly status of AT&T) could sustain a laboratory like Bell Labs. For instance, the basic ideas of: (i) satellite, fiber optic, and cellular communication; (ii) Lasers; (iii) the Internet; (iv) the electret microphone; (v) digital coding, synthesis, and recognition, of speech; and (vi) Unix were all born at Bell Labs during that time. And that, obviously, is just a very tiny fraction of the achievements. So why can't that type of research environment be affordable? I believe we need someone with Baker's visionary insight to emphasize that it is in the national interest to have a laboratory of this type: a laboratory where industrial research is done in a university-like atmosphere but where, unlike at a university, the environment is rich with problems of direct interest to industry. Perhaps a national laboratory is the answer. Or perhaps a laboratory run by an industry-led consortium could be formed. In my opinion it is imperative that means be found to set up such a laboratory. I believe it is important that a certain portion of industrial research be done without the emphasis on immediate application. The exigencies of developing products and services create an atmosphere of rapid changes that may be likened to a rollercoaster ride. That can be thrilling, but it is not possible to think about the distant future while riding a rollercoaster. And some amount of relaxed thinking about the future is necessary.

WILLIAM L. KEEFAUVER

William Keefauver. (Reprinted with permission of William Keefauver)

Gettysburg Beginnings

Gettysburg, Pennsylvania has fame far beyond its small-town nature but in my early years it was simply my hometown. I was born there in 1924 to parents who were the first in their families to be schooled beyond the 8th grade, and who became educators themselves. My mother had been a music teacher and my father a school principal who had become the Gettysburg Superintendent of Schools. Growing up during the great depression meant that one had to create one's own amusements, but with the national park surrounding the town, I had a large playground. Also, I had two cousins living on a farm just a few miles away and I frequently visited and stayed with them. My early years gave no discernible hint as to what my eventual career path would be. History and music were my main pursuits. Music was probably my major interest and the major conservatories were all I thought of as college choices. But it did not turn out that way.

I was a serial reader from an early age, with history my main focus. This was perhaps stimulated by the history associated with the town. But also, in his position my father received numerous history book samples, all of which I read. Two singular events helped make that history real. In 1938, there was a large celebration marking the 75th anniversary of the famous battle there. Over 200 Civil War veterans attended, most of them former drummer boys or the like. However, it was eerie to move among them and to imagine what they had experienced. The other was meeting a lady who came in the late 1930s to visit my father and who identified herself as the widow of

Civil War General Longstreet, She had come to insist that he remove all history books which blamed her husband for the Confederates' loss on July 3, 1863. That was the conventional wisdom at the time, so my father refused. However, in recent years, it has become well established that the fault was Lee's and not Longstreet's. We confirmed later that this woman really was Longstreet's widow. She was a teenager hired to care for him in his old age, and married him when he was in his 80s.

My early schooling was unremarkable. History and mathematics were my favorite subjects. In high school, I participated in music, dramatics, debating, and also the school newspaper. I liked sports – particularly basketball – but had no particular talent. In hindsight, my most useful activity was the newspaper, of which I became editor. This activity prompted me to learn to type, a skill not only useful in college but also much later with the advent of computers.

My music 'career' began with piano at age four with my mother as teacher. Practicing was a difficult discipline for me to learn and I eventually took up the trombone after a brief stint on drums. I learned to practice four hours a day, entered contests and one year came in second in the state. I also taught myself how to play other instruments, primarily brass, and I played in numerous town bands in the area. All this, I felt, would help me gain admission to one of the better conservatories. Discussions with my music buddies focused on the relative merits of Eastman, Julliard, Curtis and Peabody. My father however derailed this view and convinced me that music could be a great hobby but was not a wise choice for me as a career. Although I was not at the time pleased with this advice, I gave him sincere thanks years later.

My college options were very limited. Although Superintendent of Schools was a grand title, school-teachers and administrators at that time were paid very little. Thus, I could live at home and go to Gettysburg College, or, like my older sister, go to Penn State where tuition for in-state students was then quite low. The decision became Penn State when I entered and won an essay contest, which carried with it a Penn State scholarship in the munificent sum of $100 per year. In today's climate this seems paltry but it brought my tuition down to $39 per semester, including lab fees and football tickets! College guidance had not yet been invented, at least in Gettysburg, so I was on my own in picking a major. I rather liked physics and did not see math as a career choice, so entered 'physics' on my college application. Because of a chance conversation with my former Boy Scout leader who was already in college, I switched to electrical engineering (EE).

Penn State had about 5,000 students at that time – quite an increase over my high school, which had less than 500. However, there were only twenty of us in EE. I almost changed course again when I learned that all of them were either ham radio operators or had built all kinds of electrical things, none of which I had done. But I did not change and surprised both my father and myself by making the dean's list the first semester. More important to me at the time, however, I was one of only a few freshmen to be accepted in the college marching band as well as the symphony orchestra and a student jazz band. Thus, I could continue my interest in music.

Pearl Harbor

My freshman year began in the fall of 1941, three months before Pearl Harbor. The college quickly switched to an accelerated program, which shortened semesters so as to permit students to graduate before likely entering military service. This meant that I reached my junior year while still only eighteen years old. Many of my classmates were enlisting, and I decided that I should do the same even though engineering students were being deferred from the draft.

One day in February 1943, I spotted a notice in the college newspaper that a recruiter for meteorology cadets for the Army Air Corps would be speaking that evening. I went, found it interesting, and signed up. A few weeks later, I was called to active duty and was soon studying a new subject. After an intense nine months of college-level material, I received my commission and soon became base weather officer in Grenada, Mississippi at age nineteen.

Just as I was getting accustomed to my new environment and career, the well-known foibles of the military asserted themselves. What were really needed, it had been discovered, was not meteorologists but communications officers. After a brief orientation, I received tropical gear for an assignment to the 'hump' between India and Burma. But the orders that came through said "Greenland" and that is where I spent the next year and a half – with different gear – in communications and in cryptography, leading to a life-long interest in the latter.

There I saw much of the 8th Air Force pass through first eastbound and then, after VE day a year later, westbound. I was on duty in the crypto office the morning we received a lengthy top-secret message signaling that the invasion of Europe was about to begin.

William Keefauver in the service. (Reprinted with permission of William Keefauver.)

Several months after returning to the States, I was noticed for discharge, which put several major decisions in front of me. I could take a posting to Hawaii, with a promotion, and be eligible to retire at age thirty nine. Or, I could enroll at the University of Chicago and, in one semester, with credits for my military meteorology study, get a degree in that subject. Lastly, I could return to Penn State for two years to complete my EE degree. With persuasion from my jazz friends, I choose the last option.

With a three year hiatus in college work, I was somewhat apprehensive but managed to make the dean's list most of my remaining semesters, graduating in February 1948. During these final two years, the GI Bill paid for tuition and books plus a monthly stipend that covered room and board. As a result, I was earning enough playing jazz to send money home.

As great as my many experiences at Penn State were, clearly the greatest was meeting and dating a young out-of-state student from New York City, named Barbara Atkins. We hit it off from the very beginning and became engaged in December 1947, one month before my graduation and one year before hers. Had it not been for my WWII hiatus, I would have graduated before she entered and we would likely not have met. Our marriage was on July 9, 1949.

Bell Telephone Laboratories Calls

As I neared graduation, I realized I had another major decision to make. What did I *really* want to do? With my WWII meteorology experience, plus a degree in EE, a career in the infant but rapidly growing airline industry seemed attractive. An MBA was also considered. As much as I enjoyed the more theoretical aspects of EE, I could not see myself working at what I perceived an electrical engineer would be doing. But I also had law in the back of my mind. My father had long thought of becoming a lawyer and had a collection of law books many of which I had sampled and found interesting. With the GI bill, I could have gone to day school but, at the 'advanced' age of nearly twenty four, I felt I should get to work.

Thus, I responded to a notice that a recruiter from Bell Telephone Laboratories was seeking candidates for its patent division. Those accepted would begin full-time employment but would be required to work toward obtaining a law degree. This could be done at any one of several schools not far from the Labs' New York City location, all of which had evening divisions and permitted one to get a degree in four years rather than the three required for a day student. Although I knew nothing of patent work, it was emphasized that it was technology intensive, which appealed to me.

Moreover, there was the opportunity to become affiliated with the most prestigious laboratory of its kind in the world. I was aware of its reputation from a senior level textbook, which was replete with footnotes citing Bell Labs authors. When I received an offer of employment, it was too good to pass up and I accepted without hesitation. I began employment one week after graduation. At the same time, I enrolled as a student at the New York University School of Law, beginning evening studies there the following September. With a room in Greenwich Village, I was within walking distance of both my Bell Labs office and also the law school on Washington Square.

I had been assured that this arrangement would last but, not unlike the army, plans changed and less than two years later, I was transferred to Murray Hill, New Jersey. This meant an hour and a half commute both mornings and late afternoon. Classes ran ten hours a week with some going to 10:00 pm. To manage time, I set aside all day Saturday for studies and took off Sundays to relax. I graduated from law school in 1952 and admitted to the New York bar in the spring of 1953. I was subsequently admitted to other federal bars including the Court of Appeals for the Federal Circuit, which hears appeals in patent cases and also to the United States Supreme Court, the ultimate arbiter of patent disputes. Now I was a fully qualified 'patent attorney' although I always considered myself an attorney who specialized in patent law.

When I made my initial decision to join Bell Labs, I anticipated working there until I got my law degree and then going elsewhere. However, in the years to come it turned out that I only once considered making a change and that was to accept a position offered me at Motorola shortly after completing law school. But, without too much agony, it was rejected. I had already found Bell Labs an exciting place to be.

Patent Attorney— Engineering Plus Law

From a technology perspective, my timing could not have been better. Less than two months before my arrival at Bell Labs, three of its physicists had invented the transistor. This would not be announced until June of 1948 but was quickly seen as a seminal invention that would revolutionize electronics. What was not seen was its long term impact on society at large.

One of my early assignments was to a small group formed to identify patentable inventions as applications of the transistor were being explored. This was an exciting opportunity, which brought me into contact with some of Bell Labs' brightest engineers. From this experience and many others in the years to come, I learned that most, and nearly all, top-flight engineers and scientists are teachers at heart. No matter how many questions I had – and I had plenty – I could find an expert who would take the time to patiently explain what was puzzling me.

And, at Bell Labs, these teachers were frequently world-renowned in their specialty.

The inherent nature of patent work meant that I was working on the forefront of technological change. This made continuous education a must. While much of it came from day-to-day work, instruction was also available from an endless series of in-house seminars. Usually, the participants were the innovators who were driving the change. Often there were outside speakers, particularly in research topic areas. I recall, for instance, an early seminar on computing led by John von Neumann of Princeton University.

Some of my early work included logic circuits, which several times led to conflicts with Sperry Corporation. Sperry had acquired the ENIAC technology developed at the University of Pennsylvania. These pioneering systems were the very first electronic computers. Sperry tried to commercialize this technology but soon lost out to IBM, once again vividly proving that simply being first with new technology does not guarantee commercial success.

I also learned from this work that George Stibitz of Bell Labs had preceded ENIAC with a complex calculator constructed with telephone switching equipment and which could be programmed to solve complex mathematical equations. Although not electronic like ENIAC, which used vacuum tubes, it had substantially the same functionality and was programmable by a punched paper tape. When

Bell Labs showed little interest in his work, Stibitz left to become a professor at Dartmouth. I had the pleasure of meeting him years later when Bill Baker became interested in publicizing his seminal work and arranged an oral history interview with him.

My cryptographic background, and the security clearances required to work in that area, got me into some fascinating technology not known to most employees. Some of it related to Bell Labs' WWII work for which patent applications had been filed but were still subject to governmentally imposed secrecy orders. The Labs, based on its early work on Vocoders, had developed a machine that permitted President Roosevelt to talk with Churchill. A predecessor of this machine had been demonstrated at the 1939 World's Fair in New York City and had been modified to provide the necessary privacy. This work began a long relationship between Bell Labs and what eventually became the National Security Agency (NSA). The long term value of the Labs' contributions is shown by the fact that the government did not lift many of these secrecy orders until decades after the war.

My early patent work also got me heavily involved with what became known as digital transmission. Although pulse code modulation (PCM) had been invented elsewhere, it had been developed at Bell Labs during the war as part of its classified work. This work was continued in the research organization after the war and I was lucky to be associated with it, for I found the entire field of digital multiplexing and PCM very interesting. On one occasion I wrote a patent application for W. D. Lewis, later to become president of Lehigh University, and John Pierce one of Bell Labs' most eminent scientists (who insisted on being called an 'engineer' and not a scientist). It was a proposal for a nationwide digital system, which seemed far-fetched at the time but is now, in fact, a reality.

Bell Labs was best known from its published papers and inventions, most of which came from the research organization. But this was just ten percent of the company. The other organizations performed systems engineering and development. In theory, new technology would come from researchers, systems engineers would then specify how it should be deployed, and then products to implement that deployment would be developed. In practice, this orderly sequence was frequently violated. All organizations were highly interactive and it was usually hard to draw a straight line from beginning to end. I was fortunate to have patent responsibilities in both research and development organizations.

Although Bell Labs' work was focused primarily on the Bell System, about twenty percent of its development budget was work for, and funded by, the military. Major contracts were with the Army for anti-ballistic missile (ABM) systems – the *Nike* family of missiles – and with the Navy for submarine location systems (SONAR). Following a promotion in 1955, I had primary responsibility for the patent work in this area, much of which required me to get ever higher levels of security clearance. In my later years, I also had responsibilities at Sandia, a government-owned nuclear systems laboratory in New Mexico, which was managed by us under a dollar-a-year contract, first with the Atomic Energy Commission, and later with the Department of Energy. One of my last visits there prior to retirement was with a group including the head of the CIA, who was seeking technology to aid in verifying Soviet compliance with nuclear disarmament agreements.

Internally circulated technical memoranda were an important source of information for us in the patent staff and also for promoting useful interactions among the members of the technical staff. As patent attorneys, however, we could not rely on such memoranda being written in a timely manner and had to create and nurture personal contacts with our client organizations. This frequently got us in on the ground floor of new science and new developments, which made the work particularly exciting. Such interactions were not always internal since the research organization had long-standing ties with the university community. These ties were enhanced by temporary employment of post-doctoral students. These external interactions often provided challenges to the patent attorney. In case a patentable invention resulted, it was necessary to determine inventorship and get legal rights assigned accordingly. While such inventions were not the rule when conducting basic research, when they did occur, they were likely to be of considerable importance. One significant example was the laser, co-invented by Arthur Schawlow of Bell Labs and his brother-in-law Charles Townes, then a professor at Columbia University.

One of my favorite Bell Labs anecdotes concerns an informal study made by the head patent department administrator early in my career. He was intrigued by the large difference in the number of patents granted to a relatively small number of members of the technical staff and discovered that the one thing many of them had in common was that they all had lunch with Harry Nyquist. Nyquist was an outstanding engineer and well-known for his exposition of the basic sampling theorem which became known as the Nyquist interval. Like many other eminent Bell Labs innovators, he was willing to listen to other people's technical problems and was very adept at suggesting paths to follow for a solution. John Pierce was a similar person although, instead of having lunch, he would walk into a lab, quickly assess what was going on, and then just as quickly make a few comments, which would often redirect the work on a more useful trajectory. I had the pleasure, and the challenge, of writing several patent applications for John.

The availability of these experts, willing to discuss problems with their colleagues, was to my mind a major key to many of the successes Bell Labs achieved over the years.

Computer Program Patents

As a unit of the Bell System, Bell Labs focused primarily on the needs of the operating telephone companies. Examples included telephone switching systems, transmission systems for long and short haul, and terminal equipment such as telephones. Much of my early work consisted of writing and prosecuting patent applications related to transmission systems and to military work. In 1955 I was promoted and my time became more supervisory and administrative. Nonetheless, I kept in close contact with the technical organizations within my realm of responsibility so as to stay abreast of technology and also to make certain that new developments were being appropriately considered for patent protection.

In the mid-1960s, I became aware from research briefings of a new development, which seemingly was of major importance: computer programming. Knowing nothing about it, I persuaded a researcher to teach a few of us patent attorneys what it was all about. We also conducted a legal analysis and concluded it should be patentable. When we went public with this view, we faced strong opposition from IBM which then was promoting a marketing strategy which had its customers create their own programming and then make it freely available to others. We went public with our conclusion in testimony to the U.S. Patent and Trademark Office, which was struggling with a technology for which it was unprepared.

When we took an appeal from a rejection of one of our software patent applications to the courts, our initial victory was appealed by the government to the U.S. Supreme Court. Although we lost that appeal, in its decision the Court said it was not holding that computer programs were not patentable subject matter, a bit of dicta that led to later decisions upholding their patentability. Negatives are not always negative.

The importance of the issue is demonstrated by the fact that the UNIX software created by three of Bell Labs researchers was first tested in the patent division, so that we would have a full understanding of it in preparing patent applications. It was obviously of great importance and we carefully considered how to protect its intellectual property. Commercialization was considered but resisted by attorneys in the parent company because of business restrictions in a 1956 Consent Decree, which settled a 1949 government anti-trust suit. Bill Baker, Vice President, Research, was anxious to disseminate UNIX with the aim of creating a de facto standard. His proposal, which we helped implement, was to license it royalty-free to any university. This worked as he had hoped, and soon we had new employees already familiar with it. At one point, it was offered royalty free to the Digital Equipment Corp. on whose computers it had initially been implemented, but the offer was rejected.

During a visit by one of the Bell Operating Company boards, its President, Andy Smith, asked me to speak with one of his board members about a licensing problem. This member, Dr. Mary Gates, a dean at the University of Washington, told me her son Bill was forming a company and was having difficulty with our licensing people. When I called him to offer assistance, he very politely declined, saying, "You know mothers; but I can take care of myself." That was shown in subsequent years to be a monumental understatement.

My work was frequently involved with patent licensing matters. Western Electric had the corporate licensing responsibility for all AT&T entities and had hundreds of agreements based primarily on Bell Labs patents. Thus, both Bell Labs patent attorneys and technical people were frequently called on to assist the Western licensing organization to help identify important patents and to ensure appropriate coverage as technology changed. Occasionally there was litigation and we were frequently called to assist in that activity. As a regulated utility, Bell's policy was to use its patents in cross-licensing agreements to acquire rights under the patents of others rather than as a revenue source.

Broader Assignments

One interesting project I worked on was *Telstar*, the first active communications satellite, which was launched for the Bell System by NASA in 1962. This was a dramatic technological and public relations success but Congress intervened and decreed that the business should go to Comsat, a newly created quasi-governmental corporation. I also had the primary patent responsibility for Bellcomm, a company set up at the request of NASA to perform the systems engineering for the first manned space flight missions. This gave me my first contact with Ian Ross, Bellcomm's President and one of several hundred Bell Labs engineers transferred to Bellcomm to work on this project. I later had much closer contact with him when he became president of Bell Labs and I was its General Counsel.

In 1971, I became General Patent Attorney (GPA) in charge of the entire Bell Labs patent operation. In this capacity I reported to Bill Baker, then the vice president in charge of research. The patent staff included ninety or so attorneys in more than domestic locations plus patent agents in London and Tokyo. It also included a very capable support staff of several hundred people.

Being GPA brought me into more frequent and direct contact with top management. Without question, the most interesting of these contacts was with Bill Baker. In my new position, I was invited to the monthly Research meeting of the heads of the various research groups. After Bill made opening remarks, which frequently were the highlight of the meeting, the research heads would report on activities in their laboratories. These gave me a vivid picture of research at its most fundamental level. I clearly recall the month-to-month progress in the then new field of fiber optics as efforts were being made to reduce fiber loss to a level such that it would be commercially practical.

Although I had known Bill previously, I now had more frequent contact with him. He had a very wide range of interests, much of which would become apparent only over time. He also had a trait, when in a group, of speaking in riddles, seemingly to avoid revealing what was really on his mind. But, one on one, he was always very pleasant and direct.

I learned early on that he was very interested in determining the priority of new ideas. Not infrequently they would become associated with the first person to publish, and Bill wanted to make certain that the true originator was identified. He knew that our patent records contained relevant information and frequently directed questions to me. Through these and other contacts, he learned that I shared his interest in trying to shape legislation in Washington and we frequently compared notes on these efforts.

I also learned that he kept detailed records in his office of nearly everything that came to his attention and that, years later, he could and often did recover an appropriate record. Not knowing what was in these records and knowing that Bill had strong opinions on many matters other that research, this was a frequent worry during pre-trial of the 1974 government anti-trust suit. Amazingly, his files were never requested and they surfaced only when Bill retired and the many boxes holding them were removed to his home.

My Washington activities during the 1970s brought me into contact with the AT&T officer in Washington assigned to make certain that the communications needs of the White House were being met. I got to know him quite well and worked with him on some interesting projects. Many of these were classified and aimed at frustrating Soviet intelligence gathering.

My patent responsibilities also involved many classified projects for the military. One in particular stands out, since, shortly after it began, the several dozen or so of us on the project were grounded. There had been a recent spate of airplane hijackings and in view of the special sensitivity of the project, which dealt with satellite surveillance, we were ordered not to fly until further notice. Such notice did not come for nine months during which I grew accustomed to frequent train rides. When my then boss queried this change in transportation mode, I told him that at heart I was a train buff – which was true but not the reason. This was one of several projects on which I worked where traditional reporting channels were ignored because of special security requirements.

Outside Activities

My predecessor, although a very competent GPA, was strictly an internal person as far as his professional interests were concerned. On the other hand, I had become quite active in not only the Patent, Trademark and Copyright section (PTC) of the American Bar Association (ABA) but also in the Electronics Industry Association (EIA), a trade association. I chaired its patent committee and found myself working closely with the heads of companies such as Motorola and Texas Instruments on patent policy matters. I also assisted with EIA testimony before congressional committees. At the ABA, I headed a variety of committees including a software committee where we drafted resolutions supporting the patenting of computer software, which became adopted. In 1978, I served as Chairman of the entire PTC Section. These activities involved me in a variety of Congressional appearances where we tried to influence patent policy. It also gave me the opportunity to speak directly with senators and congressmen on such matters.

One of my first direct contacts was with Senator John McClellan of Arkansas, at the time the chair of the Senate subcommittee on patents, trademarks and copyrights. After asking his receptionist if I might see him, she said, "Why don't you just walk in and talk with him; he happens to be available." He could not have been more hospitable and we talked for nearly half an hour. On the strength of this experience, I frequently dropped in to chat with other Congressmen involved with patents as well as with our New Jersey representatives. Although I was not always successful, I frequently did make contact. The funding of basic research was another topic and on several occasions I provided witnesses and testimony at the request of members of the relevant Senate committee. Ian Ross and Arno Penzias were two such witnesses.

Much of this was *ultra vires* as far as the protocols of AT&T were concerned. There was in Washington a very competent organization whose sole function was to maintain contact with the executive and legislative branch. Although I eventually developed close contacts with them and frequently used their office facilities, I initially took the direct approach. In the end, however, we were all friends. I became essentially a member of that organization in 1982 while helping to lobby the defeat of HR 5158, a bill sponsored by Congressman Timothy Wirth of Colorado, which would have broken up the Bell System and Bell Labs.

As I got to know Bill Baker better, I learned that he did much the same with political leaders in the policy areas that interested him, including science, health and intelligence. When he leaned of my informal "extracurricular" activities, he frequently encouraged me not only to continue but to expand them.

During the 1970s and 1980s, I was asked to serve on and with a number of external study groups involving research and development (R/D) and also intellectual property policy. One was a commission on competitiveness appointed by President Reagan and chaired by the then chairman of Hewlett-Packard, John Young. I was not a member, but Ian Ross was, and he asked me to assist him. Frequent meetings and hearings were held about the country; I participated in most of them and was generally treated like a member since I often had to stand in for Ian, who was unable to attend many of them. On one occasion, I was asked to negotiate a contentious issue between its proponent, the President of Pfizer, and its objector, a Vice President of the AFL/CIO. Another time I was called upon to debrief David Packard, a co-founder of Hewlett- Packard and a former head of the Department of Defense, on the subject of government laboratories, the subject of a special commission he was heading. I also assisted with the drafting of his final report.

Another memorable opportunity came with a White House Conference on Productivity for which I co-chaired a study group on R/D and intellectual property policy. At one of our meetings about the country I had a stimulating luncheon conversation with a group including Judge Stephen Breyer, then a judge on the Court of Appeals and now a Justice on the U.S. Supreme Court. President Reagan attended and spoke to us at one of our meetings.

I also attended several meetings of a Presidential Commission on Education at the request of AT&T Chairman Jim Olson, who was one of its members.

Shortly after retirement I was appointed to an Advisory Commission on Patent Law Reform by the Secretary of Commerce. I had special responsibility for complex litigation issues. The Commission dealt with many divisive issues but managed to produce a series of recommendations for change most of which have by now been enacted into law.

One of the most contentious issues related to determining priority between inventors claiming the same invention by dates of conception and reduction to practice. The U.S. way for many years had been to give primacy to practice. Most of the rest of the world awarded the patent to the first to file a patent application. I had been a long time advocate of changing to this first-to-file approach, in the interest of international patent law harmonization, and frequently took part in the often heated debates on the subject in the ABA. Although opposition to this change continues, it has diminished in recent year and may yet become law.

Clearly the most enjoyable professional diversion from my day-to-day work at Bell Labs was what I refer to as my diplomatic work. In the early 1970s, I was appointed to the official US delegation for discussion of proposed revisions to the Paris Convention of 1883. This is the granddaddy of all intellectual property treaties and the developing countries claimed it frustrated their attempts at development. The delegation was led by the Commissioner of Patents and Trademarks and had two other private sector members. This treaty is administered by the World Intellectual Property Organization (WIPO), the UN organization in Geneva which hosted the discussions. I was familiar with WIPO from having taken part in an international study group there charged with drafting a model law for the protection of computer programs.

I thoroughly enjoyed these opportunities to seek common ground with representatives of the developed, the developing and the socialist – primarily Soviet – countries around the world. In the case of the Paris Convention talks, a number of preliminary meetings preceded the diplomatic conference. Also, the developed countries held separate strategy meetings prior to the Geneva discussions, usually in London, Bern, Paris or Washington.

Issues were frequently discussed in parallel meetings. Once, I served as the U.S. delegate debating issues with the Soviet representative, related to their attempt to elevate the Soviet inventor's certificate to the same status as a patent. In another, I had the fun of discussing the protection of names of wine — which I happen to enjoy — with at least seven representatives of the French wine industry who very upset with U.S. misuse of their "appellations of origin."

After the first, and unsuccessful, diplomatic conference, I had to bow out of this activity because I had inherited increasing company responsibilities and since many of these meetings were lengthy. Also, the government adopted a new policy forbidding private sector representatives to be official members on its delegations, which greatly diminished our participation. I did, however, attend one later meeting in the Hague as the representative of a nongovernmental organization.

Anti-Trust Litigation

My first taste of antitrust litigation was in the early 1950s. The Department of Justice (DOJ) had brought suit to force the Bell System to divest itself of its manufacturing entity, the Western Electric Co. My work entailed reviewing the commercialization of Bell patents, which were always somewhat suspect to the DOJ lawyers. A 1956 consent decree settled this suit without trial, permitting retention of Western Electric. It required existing patents – including the basic transistor patents – to be licensed royalty free and future patents to be licensed to any applicant on reasonable terms and rates. Since the latter happened to be our existing patent licensing policy, very little was changed. However, some of the decree's terms were to loom large several decades later since one of them restricted the company to the heavily regulated telephone business as it was then conducted.

A much bigger challenge came in early 1974, when the government filed another antitrust suit, this time seeking to break up the entire Bell System. More than forty private 'me too' suits were filed about the same time. These changed my work dramatically. Bell Labs had until this time two legal organizations: the patent organization that I headed, and an entirely separate but much smaller group of lawyers who dealt with general legal matters. Bill Baker saw that the government suit would focus in part on Bell Labs and felt that the lawyer heading its defense should be someone conversant with technology. In 1974, he had the legal organizations combined, with me in charge. This required me to establish a sizable group of lawyers, paralegals, and technical people to work on the antitrust cases. Several years later I was made Vice President and General Counsel, the first to hold that position.

Since this is not intended as a discourse on lengthy and complex litigation, I will not relate its many twists and turns. For approximately the next ten years, nearly all of my personal efforts were tied to this litigation. The day-to-day patent responsibility and what I called the general legal responsibility I delegated to very capable subordinates. My new antitrust responsibility brought me into close contact with both the lawyers in the parent company, AT&T, and also with the firm of Sidley and Austin, which had the primary litigation responsibility. Howard Trienens was a senior member of this firm and would soon come 'in-house' as Vice President and General Counsel of AT&T.

Bell Labs senior officers, primarily President Ian Ross and Executive Vice President Sol Buchsbaum were to become major witnesses, so preparing their testimony became a major work item. Although I was not the principal author, I necessarily gave its preparation close attention. As the trial date neared, much of my time was spent in Washington, with frequent trips back and forth.

As the litigation dragged out and it became apparent that the sympathies of the presiding judge, Harold Green, were not with us, Trienens involved me with a few others to explore the possibility of settlement. He later designated our settlement options as Quagmire I, Quagmire II, and Quagmire III. A major feature of each was to retain Bell Labs largely intact. It was believed that the continuation of the Labs would be seen to be in the public interest and Judge Green had seemed to understand its national importance. Nonetheless, the DOJ rejected all three proposals.

Bell Labs was funded for its research and systems engineering work by contributions from the Bell operating companies, while Western Electric funded its development work. Howard had asked me to make a presentation to senior AT&T officers on the effects of various divestiture scenarios on the funding of Bell Labs. This discussion led to a recommendation to Chairman Charles Lee Brown to accept a DOJ proposal to permit AT&T to retain its long distance operation, Western Electric and most of Bell Labs if the operating companies were divested. He and the Board decided to settle the case on this basis, explaining that a continuation of the litigation, including expected appeals would cast a cloud of uncertainty over the company for the next ten years.

The Bell Break-Up

When the Judge approved the settlement, the next big challenge was implementation. Major companies, which for a century had been closely integrated, had to be taken apart, both in terms of their physical assets and their personnel, without any loss of service. Chairman Brown likened the task to taking apart a giant airliner while keeping it flying. After less than two years of work, a plan was designed, approved by Judge Green, and implemented at last on January 1, 1984.

Although Bell Labs survived, it did not do so intact. Some research people as well as systems engineers had to be transferred to a new entity called

Bellcore, which would be owned jointly by the seven divested regional telephone companies. Also, some Bell Labs facilities had to be similarly transferred. Executive Vice President John Mayo was responsible for identifying the people and assets for transfer, and my role was primarily to make certain that legal discrimination claims would be avoided. The Bell Labs development organizations were left intact but, as we know now, this turned out to be only a temporary solution. The manufacturing facilities were spun off a decade or so later to a new company called Lucent and the Labs was further subdivided, with a portion remaining with Lucent as 'Lucent Bell Labs', a smaller portion going into a new AT&T Labs. Another decade found Lucent merging with the French company Alcatel, becoming Alcatel-Lucent. By then, "Bell Labs" was little more than a shadow of its former self.

A few years after becoming Vice President and General Counsel of Bell Labs, reporting directly to Ian Ross, who had succeeded Bill Baker as President, consolidation of AT&T subsidiaries led to my becoming also a Vice President, Law, of AT&T, with responsibility for all AT&T intellectual property law matters as well as what were once the three separate patent organizations of AT&T, Western Electric, and Bell Labs. In this capacity, I reported directly to the Vice President and General Counsel of AT&T, originally Trienens and later John Zeglis.

These dual positions continued until my mandatory retirement in 1989. Despite my AT&T-wide responsibility, I had continued to participate in all Bell Labs Council deliberations as well as numerous technical organization reviews. Keeping up with new technology continued to interest me, so these reviews were a very pleasant change of pace from antitrust and personnel issues.

As chief intellectual property lawyer of AT&T, my responsibilities included not only patents but also trademarks and copyrights. My general counsel responsibilities at Bell Labs took me into a variety of new areas of the law. Prominent was the ever-troublesome area of workplace discrimination. Others included such diverse areas as export control of technical information, tax law, bankruptcy, and workers' compensation. Fortunately, there was usually an expert elsewhere in the AT&T legal organization who could be consulted although I made myself at least conversant in each field.

Life After Bell Labs

Although company policy called my departure in March of 1989 a 'retirement', I had no intention of retiring. After briefly considering becoming 'of counsel' to a law firm, I decided to go it alone. With a computer, fax machine, FedEx, and my typing skills acquired years ago, I decided it was practical to work at home. This I did for the next fifteen years.

My first experience was to participate in a patent law symposium in Helsinki, Finland and to give a paper on the outer limits of patent law. This was quite a challenge and got me back doing legal research, which I had not had the opportunity to do for many years. It was intellectually very stimulating. I also joined the patent law speaking tour circuit, giving papers on subjects such as the patentability of computer software, secrecy orders, and also the intellectual property aspects of the upcoming integration of Europe.

Before long, I had two significant clients: my former employer AT&T and the Lee Publishing Company of Davenport, Iowa. Lee Publishing of Davenport, Iowa had one patent which they thought might be quite valuable in the communications field and I was retained to assist in this effort. This relationship lasted several years, but most of my work was with AT&T.

At the time, AT&T was involved in a massive arbitration with its divested operating telephone companies over application of the "conditional liabilities" terms of the divestiture decree of 1982. Each side chose an arbitrator and I was the AT&T choice. Judge Frederick Lacey, an outstanding retired Federal District Court judge served as the neutral. After reviewing many thousands of documents, Judge Lacey, the arbitrator chosen by the operating companies and I encouraged settlement, which, gratefully, was achieved.

At about this time, AT&T was making a play to acquire the NCR Corporation and I was asked to become one of four nominees to the NCR Board of Directors. All four of us got elected and I served in that capacity for about a year, by which time the acquisition had been accomplished.

I was retained by AT&T on two other occasions. One was to help outplace some attorneys as the company reorganized itself. The other came at the request of Chairman Bob Allen, who asked me to divide the company's roughly 10,000 patents among itself and its two entities, Lucent and NCR, which were being spun off. This took nearly six months to accomplish and would have taken longer except that Bob had made my decisions final and non-appealable.

I then began getting requests to testify as an expert in patent litigation. Not infrequently, after discussing trial dates and possible conflicts, I would learn that my initial input would be needed posthaste: in one case, the following day. Many of the cases were settled before trial and in most cases I would never leave home, thanks to my home office equipment I mentioned above. Some cases involved patent validity issues, some license contract interpretation, others the amount of a reasonable royalty. Many companies both large and small were involved. In one case, which became a major effort, I was retained as a royalty-rate witness on behalf of Motorola. This case did go to trial and became the only one of the many cases on which I worked in which I actually testified in court. In another case, I was scheduled to testify in bankruptcy court in a dispute over one month of royalties. While I was sitting in court waiting to testify, the judge appeared to tell us the parties had settled. A similar event took place in another case for which I was called as a witness. After an early morning flight to Rochester, NY, and some trial preparation, I was in court in the witness chair when the judge decided he did not want to hear what I had to say and I was dismissed without saying anything.

The final case on which I worked achieved a measure of national prominence. In my EIA role, related earlier, my committee had drafted what became EIA policy on the interrelationship of patents and standards. This was the major issue in this case since the party on whose behalf I was retained, Rambus (a technology licensing company), was claimed to have abused the standards process with its patents. The complete history of this case is yet to be written but, when it is, it will require volumes. Once again, FedEx made frequent visits to my home since my work required review of many thousands of documents.

This post-corporate retirement work was very rewarding professionally and free of the personnel concerns arising from corporate work. A major benefit was the opportunity to work with young people in the many law firms with which I became associated. Not all of my post-Bell Labs' work was patent related, however. On one occasion I was sent to Bulgaria with two law professors to serve on a panel funded by the US government. The purpose of the panel was to critique a proposed revision to Bulgaria's Soviet style copyright law, and was part of a broader effort to assist former communist bloc countries to revise their laws to help transition to market economies. Another time I was retained by WIPO to work in Geneva with a Swiss professor and a Japanese lawyer to draft a model law on unfair competition.

Probably my most interesting professional association activity was with AIPPI (International Association for the Protection of Intellectual Property) an organization with members in more than sixty countries. When originally formed, its first activity led to the 1883 Paris Convention mentioned earlier, and its main objective was to promote harmonization of the many disparate national patent and trademark laws. In the 1990s, I served as chair of the AIPPI policy committee, which had fifteen members from fifteen countries. A major task was to modernize the AIPPI organization. We usually had an annual committee meeting in London, which invariably was attended by all fifteen members. (Try to match that with a small committee in your hometown.) AIPPI's general membership meetings were held in cities around the world, so frequent foreign travel was involved. At these meetings, current topics were debated in an attempt to achieve common positions that were then lobbied with WIPO in Geneva and the European Patent Office in Munich. I later became Treasurer General and a member of the management group of officers, known as the Bureau. With both Bureau meetings – often at the home office in Zurich – plus membership meetings, two or three trips abroad each year was normal. Barbara always came with me since we both enjoyed traveling and we often tacked on vacation travel as well. This AIPPI activity gave me a much better and broader insight into the intellectual property laws around the world and also gave rise to many friendships with people in other countries.

Nor were my outside activities limited only to legal affairs. I have served as vice president of the board of the McCarter Theatre, in Princeton, New Jersey, on a board in the Engineering College of Penn State, and on the Board of Trustees of the Franklin Pierce Law Center in Concord, New Hampshire. I have also served as a school board president and as chair of the local Board of Adjustment which rules on building variance requests. I have been recognized by Penn State as an Alumni Fellow and by the AIPPI as a Member of Honor.

But, the best I have saved for last. And this is my wonderful fifty-seven year marriage to Barbara. We first lived in Greenwich Village in Manhattan. This was convenient to all of the cultural activities of New York City in which we both were interested, particularly the Metropolitan Opera. Barbara, having lived in "the city," had personal experience with its opera, theatres, and museums, but they were mostly new to me, although I was a fan of the Met through its Saturday afternoon broadcasts. So we began what became a lifetime practice of attending opera performances and visiting art museums whenever and wherever we could. The year I finished law school, we decided to take a break and took six weeks off from our employment to travel in Europe. This gave us another life-long addiction: travel. We were also blessed by the addition of children to our family. Our son Bruce was born in New York City and our daughter Betsy in New Jersey after moved there to reduce my commuting time. Sadly, Barbara was diagnosed with cancer in 2004 and succumbed in 2006.

Last Thoughts on an Exciting Period

I continue to get frequent questions on my feelings about the fates that have befallen AT&T and Bell Labs. My initial response is that I feel very fortunate to have been associated with these two companies during the years that I was. Also, while difficult to believe that the original AT&T has been superseded by an entity that initially shared only its name, it is gradually coming closer and closer to replicating its original self, and may someday do so.

But rationalizing what has happened to Bell Labs is much more difficult. It was not just another communications company or just another laboratory. From my role as Bill Baker's unofficial historian, I became very well acquainted with the great people who had both the vision and the capability to construct what I truly believe to have been the greatest of its kind. The founders of AT&T early on recognized the importance of research and began hiring Ph.D.s in physics and mathematics shortly after the first Bell Company was formed. Their documented understanding of the breadth of technology required was way ahead of its time and led to what, in 1925, became Bell Labs.

The transistor invention did not just happen. Nearly a decade ahead of its December 1947 invention, the research director realized that the vacuum tube would run out of steam eventually and that a replacement should be sought within a solid material. A specially picked team was assembled and successfully conducted that search. Numerous similar examples could be cited, though perhaps not as extreme an impact on the world as the transistor.

Perhaps it is simply the case that, for its time, a Bell Labs was the perfect solution for the unknowns that lay ahead and that, today, a different model is the answer. But I feel privileged to have been there during a very exciting period of communications technology growth and to have rubbed shoulders with those who made it happen.

For many years I have told others, only somewhat facetiously, that I had not yet made up my mind exactly what I wanted to do. In hindsight, I know that I could not have made better choices. Combining engineering and law and working for Bell Labs was, for me, just right.

EDWARD. E. ZAJAC

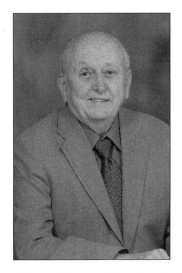

Edward Zajac. (Reprinted with permission of Thomas Venklasen Photography.)

Early Years: Cleveland, the Army, and Graduate School

In my boyhood, the view from my kitchen window was of the green, onion-shaped domes of St. Theodosius Russian Orthodox Cathedral in Cleveland, the church that was featured in the film *The Deer Hunter.* The view from my backyard was more pedestrian — black smoke belching from the steel mills of the Cuyahoga River valley.

My neighborhood was called the "South Side." Now it enjoys the name of the "Tremont District." This befits the young, upwardly mobile folks who displaced the Polish and Ukrainian immigrant families of factory workers (including my family) who lived in it during my elementary and high school days. Its big advantage, then and now, is its closeness to downtown Cleveland – about ten minutes by public transportation.

The Cleveland public schools gave me an excellent education. I was valedictorian of my class at West Technical High School, an urban vocational school. Few graduates went to college; in those days (the 1930s and early 1940s), they couldn't afford it. After high school they were expected to get a job to help with family finances. No guidance counselor took me aside to recommend I try for a college scholarship somewhere. Besides, WWII was still going on, and the draft was lurking in the background.

In February 1945, about a month after high school graduation and while awaiting "Greetings from the President of the United States," I read in the newspaper that the Case School of Applied Science (now Case Western Reserve University) was offering a competitive, full-day examination, the prize being a half-tuition scholarship ($175/year). In spite of the lurking draft, I decided to sit for the exam, and I won. The first day of class at Case coincided with the day I was to report for my draft physical for the army. I expected to fail it because a couple of months earlier, I had failed the physical for the elite navy V-12 program. The examining navy doctors heard a heart murmur. The V-12 program would have sent me to college at the navy's expense to get a bachelor's degree and a navy commission as an officer.

Unlike the navy doctors, the army doctors couldn't hear a heart murmur. They admitted they could hear something, but that what they heard was "purely functional," and they suggested that I was in great shape and qualified for the most physically demanding service. Off I went to Camp (now Fort) Hood, Texas for both basic training and special training as a radio operator in the Army Tank Destroyer Corps.

I was discharged after twenty six months of army service, the last eight in Germany in the Counter Intelligence Corps. As my discharge date neared, I applied from Germany to Cornell University and only Cornell. The reason was that two guys in my barracks in basic training had completed their freshman year there and incessantly talked about how great it was. I decided that it would be broadening to go to college somewhere other than in Cleveland, an opportunity I now had because of the GI Bill. I was too naive to realize that Cornell might turn me down. Thank God, it didn't. I graduated first in my class in mechanical engineering, in June 1950.

Upon graduating from Cornell, I applied to about a dozen schools for graduate work and received fellowship offers from all of them. Princeton's was financially the most lucrative and I chose it. The consequences taught me that finances should not have been the deciding factor. Attending a few classes revealed that, compared to Cornell, Princeton had a mediocre engineering program, with mechanical engineering, my major, perhaps the most mediocre of all (this was rectified a few years later when a new administration lifted Princeton's engineering program to the level of its other programs). It also caused me to more carefully investigate graduate programs before I enrolled in one and to change my major to engineering mechanics, the mathematical core of mechanical, civil, aeronautical, and other engineering specialties. The investigation also led me to Stanford, which, at the time, had the best engineering mechanics program in the U.S.

Getting a Job: Bell Labs

As my Ph.D. training was coming to an end, I had to decide what to do next. In the long run, I was sure I wanted an academic career. But, being an engineer, it seemed silly to teach at the university level without any prior practical experience. It was a terrific buyers' market for engineers and a job interview was almost guaranteed to result in a job offer.

I explored about a dozen firms. Bell Labs was in a class by itself. My dream came true when Hendrick Bode, director of the Bell Labs Mathematics Research Department, offered me a job. In doing so, he spoke in his usual way – in a corner, talking to the wall in circumlocutions. I finally figured out that I was being hired on probation. If I flunked, I would be sent to the mathematics research center in the division doing military research.

I considered myself a hot shot because of all the honors I had received on the way to a Ph.D. But then I was reminded of the all-conference, all-American, all-everything end we had when I was a graduate student at Stanford. When taking the football team picture at Stanford, the photographer always placed him in the back row with the big guys. When he got to the NFL, a sports writer asked him what the difference was between college and NFL football. He said that, in the pros, unlike in college, the photographer always placed him in the front row with the little guys. At Bell Labs, I quickly realized that I was now a front row guy. The honors I had received in getting my PhD were surpassed by most of members of the Mathematics Research Department. A typical member's history was: graduation from high school at sixteen, bachelor's degree at nineteen, and PhD at twenty two, while I was an ancient twenty eight when I got my PhD.

Research on the Laying and Recovery of Submarine Cable

Sergei Schelkunoff, my boss, arranged for me to meet with the engineers who were developing the second-generation telephone submarine cable. Although the first transatlantic *telegraph* cable was laid in 1857 and 1858, the first transatlantic *telephone* cable had been laid only in 1956, in two stages, using one cable for east to west transmission and a second, separate cable for west to east. The technology used to make and lay the cable was a minor modification of submarine *telegraph* technology, one that evolved from the first, 1857 transatlantic telegraph cable. By the mid-1950s, it was antiquated. One group of Bell Labs engineers was working on the next generation telephone cable, which would use more advanced technology, while still another group was researching and developing a subsequent generation using still more advanced technology. The result of the analysis that I did for the submarine cable group was a 78-page paper in the Bell System Technical Journal, "Dynamics and Kinematics of the Laying and Recovery of Submarine Cable," published in September 1957. The submarine cable engineers praised it to my bosses in the Math Department. My being transferred to military research was not mentioned again.

Computer Animation

In the early 1960s, Bell decided to try out the John Pierce/Arthur Clarke idea of communication via satellite. The first experiment was the *Echo* satellite – a passive, Mylar balloon reflector. Signals sent to it bounced back to earth. Then came *Telstar*, an active satellite, with equipment on board that could not only reassemble and amplify the signal it received, but could also stabilize the satellite's orientation so that one face of the satellite always pointed toward the earth. *Telstar* was also an experiment and not meant to be used for the long term. A couple of Bell Labs groups started research and development on its successor. I was part of a group that, together with personnel at MIT's Draper Laboratory, worked on a design that used two gyroscopes with spin axes in a V-configuration to stabilize the satellite's orientation ("attitude").

It takes three numbers to specify the orientation of a rigid body. The output of a computer simulation of the motion of a communications satellite thus consisted of three graphs showing these three numbers as a function of time. It is not easy for a human being to visualize the motion of a satellite from staring at three such graphs. It came to me that instead of showing graphs of these three numbers versus time, why not represent the satellite as a domino with plus signs substituting for the domino's dots? A perspective drawing of the domino would be trivial to compute, as would a series of such drawings, one for each instant of time. In other words, why not compute a motion picture of the orientation of the domino versus time? I explored the idea and found it easy to implement. The result was a film, "Simulation of a Two-Gyro Gravity-Gradient Attitude Control System." It was one of the first computer-animated films.

The Beginning of Legal and Regulatory Games

In the mid 1960s, Western Union had about 35,000 subscribers to Telex, its telegraph network, while Bell had about 55,000 subscribers to TWX, its competing service. Western Union complained to the Federal Communications Commission that Bell's telegraph prices were predatory and that Bell was using its non-telegraph services to subsidize its telegraph services, thereby making possible Bell prices that were lower than Western Union's.

The FCC chose not to directly deal with Western Union's complaint. Instead it expanded it, announcing that it would mount a comprehensive investigation into Bell's pricing policies for *all* of its communication services, not just its telegraph services – the first such investigation since the 1940's. In return, AT&T created a task force of about a dozen people to formulate a strategy on how best to respond to the FCC. The task force's membership had a Noah's Ark design, with one member each from all relevant disciplines – one lawyer, one communications engineer, one regulation specialist, etc. I was appointed as the resident mathematician. I never asked why the AT&T folks thought the task force needed a mathematician.

Both sides hired outside economists as consultants and expert witnesses in FCC hearings. The FCC did something unprecedented. It hired an outside *econometrician*, Myron J. Gordon of Rochester University, to construct an econometric model that would yield the prices that AT&T should charge. AT&T was caught flat-footed. It had no corresponding econometric model of its own, in spite of having world-class specialists in almost every discipline *except* economics and econometrics. The best it could do was to use John Tukey, an eminent statistician, who had appointments both at Bell Labs and at Princeton University. Top AT&T executives were heard to complain "Where's our offense? All we have is a defense." The result was that I was appointed the head of a new, Bell Labs economics research group and tasked to staff it.

A New World Class Economics Research Group is Born

The effort worked out well. Within a few years we had a group of about two dozen PhD economists that was considered to be on a par with the top university economics departments. We tried to preempt the parts of economics that might be used by opposition economists giving anti-Bell expert testimony. The idea was that any economist contemplating testifying against Bell would find that we were there first, and had published seminal papers in top academic journals on the very subject of his testimony. Thus, we were well-prepared not only to carefully analyze it and discover any errors or contradictions in it, but we could also introduce testimony of our own that was favorable to Bell or at least neutral. We now had an offense at the ready. Bell never had to use it. The FCC stopped hiring academic economists to give negative testimony against Bell.

Entering Academia

On January 8, 1982, Bell and the U.S. Government agreed to settle a pending federal anti-trust suit against Bell. The settlement, called the Modified Final Judgment (MFJ) broke up the Bell System. Western Electric, Bell's manufacturing arm, became Lucent, a new, separate corporation. Bellcore, a new research laboratory, was spun off from Bell Laboratories, and Bell's twenty three operating companies – Ohio Bell, Indiana Bell, etc., – were grouped into seven separate, independent corporations, the "Regional Bells." The new entities were to become official on January 1,

1984 – almost two years after the MFJ. Those two intervening years consisted of chaotic bargaining as to how the assets and personnel of Bell Laboratories and Western Electric were to be allocated to the new entities.

By January 8, 1982, I had been eligible for retirement for several years but was still only fifty six years old. The chaos within Bell was a great motivator for me to start thinking seriously about retirement from Bell and trying something new. I started by making a trip down the West Coast to explore employment possibilities. While on the trip I received a phone call from Helmut Frank, chair of the search committee for University of Arizona's economics department, which was looking for a new department head. He invited me to Arizona to be interviewed for the job. My wife was adamant. She hated heat; under no circumstances would she move to Tucson. Helmut said, "I understand her feelings, but what have you to lose in coming to Tucson to talk to us?"

I took him up on the offer. My wife followed in a second visit and changed her mind about living in Tucson. The University offered me the chair position; I accepted and retired from Bell Labs. Off we went to the southwest to live sixty miles north of the Mexican border, in an area rich in Hispanic traditions and culture. It was the smartest thing we ever did. The University of Arizona's economics department was the world leader in experimental economics, thanks primarily to the outstanding work of Vernon Smith, work that, in 2002, earned him the Nobel Prize in economics. Vernon attracted other outstanding economists and students, who in turn took academic jobs and expanded experimental economics as a discipline. There is possibly nothing more exciting in academia than being in on the ground floor of a new academic discipline. I, of course, was not one of the founders of experimental economics, but it was nonetheless exciting to be an observer of the unfolding and rise to eminence of the department's main research activity.

I stepped down as economics department head in 1991 to do full-time research and teaching. A sabbatical year in 1991-92 allowed me to start a new book, *Political Economy of Fairness*, published by the MIT Press in 1995. I have in mind another book, *Political Fairness Games*, aimed at a lay readership. Despite these academic experiences, however, I look back at my Bell Lab years as formative and wonderful.

WILLIAM OLIVER BAKER

William O. Baker. (Reprinted with permission of Alcatel-Lucent USA Inc.)

Note: This condensed biography of Dr. William O. Baker was written by A. Michael Noll, based on materials in Dr. Baker's files and on interviews with Dr. Baker conducted during the summer of 2002. A lengthier version of this biography is posted on the Baker web site at www.williamobaker.org. (© 2009 A. Michael Noll)

From Turkeys to Tigers: Quaker Neck to Princeton University

Born in 1915, William Oliver Baker was the only child of Harold Baker (1870-1954) and Helen May Baker (1881-1945; nee Stokes). His parents were from Brooklyn, New York, and his father Harold worked from an office on Wall Street as a commercial agent of the Central Vermont Railway Company and Grand Trunk Railway System. They married in 1912, and in 1913 moved to a 235-acre farm (the Comegys Bight Plantation) on the eastern shore of the Chesapeake Bay in Quaker Neck, Maryland. Their son William was born in the farmhouse there in 1915, and William grew up helping his mother with the duties of running the farm. A noted authority on animal husbandry and for a time the premier turkey breeder in the United States, his mother experimented with methods to raise healthier turkeys, even writing a definitive book on the subject. His father, too, possessed an inquisitive nature and added an interest in mining to his work in railroad traffic analysis. It is likely his mother's use of chemicals to combat parasites in turkeys and his father's collection of minerals influenced the young Baker's interest in materials science and chemistry. In 1937, his parents sold the farm and moved to New Jersey, presumably to be closer to William, who was then a doctoral student at Princeton University.

Helen M. Baker clearly had a considerable influence on her son. Growing large flocks of turkeys had been a challenge because of their fragile nature. However, Mrs. Baker solved these problems through careful study and experimentation, and her pioneering efforts are one reason why turkeys today are so affordable and available. In 1927, she raised one thousand turkeys (the largest flock ever) and sold them for $15,000, as reported in the *Turkey Tribune*, November 1928. In 1930, she raised an unprecedented flock of 1,900 turkeys, and in 1932, a flock of 2,800 turkeys. These were tremendous accomplishments for which she achieved coast-to-coast fame with articles describing her work in such publications as the *Rural New Yorker*, the *Washington Post Magazine*, the *Baltimore Sun*, and newspapers as far away as Pueblo, Colorado and San Jose, California. Helen Baker was known as "the Turkey Lady" and her letterhead stated "Maryland Turkey Farm - originators of Baker's Bronze Beauties," as her brand of turkeys was known. Her Baker's Bronze Beauties were predominately sold as breeding stock, at prices from $15 to $100 in 1928.

Helen Baker frequently gave public addresses and wrote articles describing how to successfully raise turkeys. In 1928, she published her book on turkey husbandry, which went through two editions by 1933, with the second "dedicated to my Beloved Son William Oliver Baker whose sterling qualities are my inspiration."

Until 1927, Dr. Baker attended a one-room, one-teacher, grade school (Quaker Neck Elementary Public School), headed by Mrs. Edna E. Faulkner, with about two dozen other students from the first through the seventh grades. He then attended Chestertown High School, graduating in June 1931 at the tender age of fifteen.

Baker entered Washington College in Chestertown, Maryland, in the fall of 1931, commuting from the farm in Quaker Neck, some ten miles away. There, he was editor of the *Washington Elm* student newspaper, a Shakespeare Player, and a member of the Dean's Cabinet, the Debating Society, the Chemical Society, the Honor Society, and the Silver Pentagon Society, among other activities. He was nearly a straight-A student there and was awarded the Visitors and Governors Scholarship for Men for the 1934-35 academic year. Dr. Baker earned his B.S. in physical chemistry in 1935, graduating Maxima Cum Laude and giving the valedictory address. Later that year, he entered Princeton University.

Dr. Baker received a Ph.D. in physical chemistry from Princeton University in 1939, graduating Summa Cum Laude, and completing his studies in a little more than three years. At Princeton, he studied the electrical properties of molecular crystals, a field which later became solid-state physics, and performed research, with Prof. Charles Phelps Smyth. Dr. Baker was named both a Harvard University fellow (1937-38) and a Procter fellow (1938-39) for his research and studies at Princeton.

Bird watching, boating, and hunting were some of his favorite pastimes from his youth. His stint at Princeton first brought him to New Jersey, where he would spend a great deal of time hiking in a natural preserve today known as the Great Swamp. In the 1950s he would go woodcock hunting in the marshes along the Passaic River with David (Duke) Dorsi, a long-time glassblower at Bell Labs (see Chapter 6).

Bell Labs: From Researcher to Chairman

Dr. Baker joined the Labs in May 1939 as a member of technical staff, working initially in a converted garage in Summit, New Jersey. His early work at Bell Labs focused on synthetic rubber, and he discovered that microgel networks formed, with unfavorable properties, if the processing was not properly controlled. Dr. Baker's unique talents as a young chemist brought him swiftly to the attention of senior management and he was quickly promoted through the ranks to become Head of the Polymer Research and Development Department at Bell Telephone Laboratories in 1948. From 1951 to 1955, he was Assistant Director of Chemical and Metallurgical Research. After a short period as Director of Physical Sciences Research, Baker followed in the footsteps of his mentor and predecessor, Ralph Bown, with whom Baker sailed summers at Cape Cod to become Vice President, Research, in 1955.

As Bell Labs' research executive, Dr. Baker was a dedicated defender of long-term research. Well aware of the pressure to produce practical short-term results, he recognized that practical problems stimulated very successful research. Yet he realized there had to be the freedom to investigate openly and sometimes even to fail. Dr. Baker deplored attempts to make basic scientists become design engineers. Paraphrasing his words, he once stated that such an over-compression of the span between discovery and use would cause the whole system to crumble "from internal pressures and implosions."

A number of important advances occurred at Bell Labs during his eighteen-year tenure there as Vice President, Research. These advances included magnetic bubble memories, the laser, charge-coupled devices for imaging, the electret microphone, and the UNIX computer operating system. The considerable effectiveness of his leadership of research at Bell Labs was the result of his early work there as a researcher, which stimulated him to think about the process of research and then develop a philosophy that he was able to implement. Testament to his influence on research is the total of eleven Nobel prizes that were given to, or were based on research done by, Bell Labs researchers during his tenure at the helm of the research organization.

Dr. Baker was issued eleven patents (on only thirteen applications) for his own inventions during his early research at Bell Labs. This research focused on the basic science of plastics, natural rubber, and other substances, emphasizing crystalline molecular structures and macromolecules. The practical motivation for his early wartime research was the search for a replacement for rubber. What he had discovered in 1943 was that the high temperatures used in the drying of synthetic rubber were creating too much gel in the final product. He was able to get manufacturers to adopt the results of his discoveries for improving the manufacturing processes, and the results of his research were essential to victory in World War II. His research into polymers was vital to the Bell System for electrical insulators, polymer carbons, and semiconductors, and was later applied to heat shields for missiles and satellites. His research and that of his colleagues enabled the replacement of lead sheathing of telephone cables with plastic, which had both a tremendous cost and weight saving as well as a positive impact on the environment.

His ascent measured in full, Baker was elected President of Bell Labs in 1973 and served until 1979 when he was elected the first Chairman of the Board. He retired as Chairman of Bell Labs in 1980, but continued his many activities in advising various foundations, academic institutions, and government agencies until his health started to fail a few years before his death on October 31, 2005.

National Influence: A Diplomat of Science

Dr. William O. Baker has been seen by many as "a diplomat of science" and "a science patriot" for his service to the United States in advocating, championing, and advancing science and technology in such national areas as the content of basic education, national security, health effects, and the environment.

Dr. Baker's considerable influence came from his ability to draw upon substantive knowledge, whether from his own research or from that of his colleagues at Bell Labs, and to apply that knowledge to problems of national significance. Accordingly, he served the role of an ex officio advisor in matters involving science and technology to Presidents Eisenhower, Kennedy, Johnson, Nixon, and Reagan.

Dr. Baker's influence on Washington projects and agencies began in 1956, on national security issues, under President Eisenhower. The "Baker Report" of 1958 had major impact on the technology of information-gathering by the intelligence community during the Cold War, including the use of special computers and satellite reconnaissance. In 1959, at the request of President Eisenhower, Dr. Baker developed the plan for the establishment of the Defense Communications

Agency, which was eventually implemented in 1961 under President Kennedy.

Presidents Eisenhower and Kennedy took Baker into particularly close confidence, including involving him in meetings with heads of state at the White House. Dr. Baker had an ability to apply substantive knowledge in an understandable fashion — as opposed to dabbling in ethereal policy — that was highly valued by the leaders of the United States.

He served formally as a member of the President's Science Advisory Committee (PSAC) and also on the President's Foreign Intelligence Advisory Board (from 1957 to 1977 and again from 1981 to 1990). He was a staunch supporter of a strong intelligence enterprise for the United States, utilizing the most advanced communications and computing technology.

Long before the personal computer, Dr. Baker was a strong advocate of the use of computers in research. He focused the attention of the academic community on this opportunity in the seminal conference "The Human Use of Computing Machines," conducted at Bell Labs in 1966. He promoted the use of computers in communication systems and libraries and has been a visionary in the then new concepts of information science and technology. He wrote and published papers that explained and advocated the use of these technologies in education and libraries.

His interests ranged into many areas other than communications science and technology. He was instrumental in stimulating investigations of the health effects and potential toxicity of various materials in use in manufacturing, and their impact on the environment. He actively promoted a systems approach to solving broad problems.

A resident of New Jersey since entering Princeton in 1935, Dr. Baker was active in promoting science and technology throughout the state. He was a charter member in 1967 of the New Jersey State Board of Higher Education and a founding member in 1985 of the New Jersey Commission on Science and Technology. Dr. Baker energetically promoted the reform of higher education and the granting of autonomy to community colleges in New Jersey.

His interest and commitment to education was national as well, and he was an active champion of educational reform nationwide. The 1983 report, "A Nation at Risk," issued by the National Commission on Excellence in Education and largely written by Dr. Baker and Prof. Gerald Holton of Harvard University, had a significant impact in identifying and focusing attention on the national issue of literacy.

Dr. Baker served as a trustee of the Mellon Foundation and, in 1965 and 1966, was a member of a small group of trustees (headed by Dr. James R. Killian, Jr.) that had been tasked by Paul Mellon to investigate coupling his Mellon Institute with the Carnegie Institute of Technology. The work of this group resulted in the merger of the two institutions in 1967 to create the Carnegie Mellon University.

In addition to the many honors he received, Dr. Baker encouraged the conferring of distinction on many others for their contributions to science and technology. He was instrumental in the honoring of AT&T Bell Laboratories in 1985 with the first Presidential Medal of Technology ever awarded to a research institution as a whole, and he later encouraged a similar honoring of the Du Pont Company.

An Assessment

Dr. Baker, as a chemist, knew the meaning of "catalyst:" namely, a substance that facilitates a chemical reaction but is itself not consumed directly. He was that catalyst for Bell Labs — and one might say for technological research in postwar America — having had considerable influence on various wide-ranging institutions and people over the many decades of his career. He facilitated and stimulated many advances in science, technology, and society. Throughout his career, he humbly eschewed personal publicity, instead exerting influence quietly and working effectively behind the scenes. He had the ability to understand and champion the research of others in a wide variety of disciplines and then to apply the results of that research to practical problems on a broad scale, both for the Bell Telephone System and for the United States as a whole.

Many with whom Dr. Baker has interacted tell stories of his personal commitment and concern. Numerous researchers at Bell Labs remember fondly how he would inconspicuously appear in the lunch line and would ask politely whether he could join them for lunch. Other Bell Labs employees in staff positions recall how he would go out of his way to wish them a happy holiday. He always expressed consideration and appreciation to all the people with whom he crossed paths, and they were many. At Bell Labs, he set a tone of openness, respect, and courtesy that permeated his entire organization.

The words that many of his friends and colleagues use in describing him mention his style and grace, his wisdom that enriched the life of the nation, his ability to lift the spirit and inspire, his tireless humanitarian efforts, and the mixture of consideration, graciousness, and moral courage that was uniquely his own. Indeed, he is particularly noted for his strong sense of the highest ethics. His own words perhaps best summarize his inspiring approach to life:

> With all its beauty and power, the age of science is in no way old enough to tell us what to do or what to think, but only sometimes what to ask.
> –William O. Baker, 1968

AFTERWORD

The contributions in this book highlight the interaction of people who made Bell labs the birth-
place of the information age. Although great talent
was gathered at all levels, it was the leadership of
Dr. William O. Baker that allowed the critical synergy
to occur. Therefore, it is a privilege to have been asked
to contribute a brief tribute to Bill Baker, with whom
I worked for over three decades, both at Princeton
University and the Andrew W. Mellon Foundation.

Most of what I have written below is based on
my close partnership with Bill Baker at the Mellon
Foundation. He was responsible for my move from
Princeton to the Foundation, and it was due in no small
part to my respect for Dr. Baker that I accepted the
presidency of the Foundation with so much optimism
about what could be accomplished. He was a superb

chairman of the board, and the Foundation profited greatly from his stewardship.

But first a brief word about my association with Bill Baker at Princeton, where he served with great distinction as a charter trustee during tumultuous years, including campus controversies over Vietnam, the adoption of coeducation, and the dramatic strengthening of the Princeton faculty, especially in the life sciences. During the Vietnam days, Bill Baker would regularly call my office to see how we were doing, and whether we were able to pursue the essential business of the University at a time of political turmoil. He had set a steady course, and he was very reassuring. He was even more important in helping the trustees of the University understand that Princeton simply had to be a leader in the life sciences if it was to fulfill its mission as a leading research university.

Given my great respect for Bill Baker, I was singularly honored when he came into my office one day in 1987 to ask if I would consider moving to the presidency of the Andrew W. Mellon Foundation. Having served for fifteen years as president of Princeton, I felt that it was time for me to "declare liberation for the faculty" and to accept new challenges at one of the country's leading private foundations. Thanks in large part to the warm, collegial, and constantly stimulating relationship I enjoyed with Bill Baker in his role as chairman, I never regretted my decision.

When it came time, in 1990, to record the Foundation's appreciation for Dr. Baker's more than twenty years of service as a trustee, not to mention sixteen years as chairman of the board of trustees, we recognized his insistence that the foundation always take a long-term approach and "stay the course."

Needless to say, it was Bill Baker's own extraordinary record of achievement in a mind-numbing range of activities that gave credence to everything that he said. Dr. Baker's entry in *Who's Who* begins by identifying him as a "research chemist." Rarely have words concealed so much – as the next ten inches of the entry illustrate. Dr. Baker was a pioneer in the organization of science and learning and the harnessing of their power to the needs of the world, through his leadership of Bell Laboratories, his service as advisor to every president of the United States from Dwight Eisenhower to Ronald Reagan, his chairmanship of the board of Rockefeller University, and his participation in countless other activities and organizations.

The extraordinary range of his accomplishments notwithstanding, it is the quality of Dr. Baker's contributions that is most noteworthy. His understated way of expressing himself ("we wonder," "we suspect," "we note") could be very deceptive. All who worked with him learned how carefully one must listen to avoid missing the absolutely central point, quite apart from the learned nuance. Dr. Baker's breadth of insight, combined with an unusual capacity for quietly effective leadership, has transformed some of the major institutions of our society. That is not too large a claim.

At the Mellon Foundation, Dr. Baker was committed from the very start to its emphasis on "higher learning," which was always seen as an instrument in support of "well-doing" and "well- being," the twin aims of the Foundation. He was more responsible than anyone else, save Paul Mellon himself, for the Foundation's evolution into a national institution with a range of programmatic objectives as detailed in its annual reports.

At the conclusion of the final meeting of the trustees chaired by Dr. Baker, Paul Mellon read a resolution, which I now quote, because it speaks so broadly to Dr. Baker's qualities.

WHEREAS William O. Baker has served more organizations engaged in a greater variety of important purposes across wider fields of knowledge and practice than is either common or indeed scarcely detectable in the sublunary worlds of ordinary mortals; and

WHEREAS no organization has received more of his devoted time, attentiveness, and commitment than has The Andrew W. Mellon Foundation; and

WHEREAS he has counseled and guided this Foundation as a trustee for two decades and as chairman of the board for 16 years, giving tangible expression to the deeply humane values that the foundation embodies; and

WHEREAS he is a clear Copernican, insofar as he has always demonstrated a distinct preference for a world that, like himself, is in perpetual motion, but motion that is regulated by laws as elegant and purposeful as they are ultimately hidden from common view; and

WHEREAS he is a devout Newtonian and figure of the Enlightenment who, with the aid of gravity, has remained firmly rooted upon this amiable planet Earth, contending with its bedeviling problems in ways that are always intellectually ingenious, syntactically inventive, and unerringly effective; and

WHEREAS he constantly approaches but is careful never to exceed the speed of light or permit Einstein's concept of relativity to slip inadvertently into any notion of mere relativism that might blur the clarity of those high standards to which he is unequivocally dedicated;

LET IT THEREFORE BE RESOLVED that William O. Baker, having been present at the creation of The Andrew W. Mellon Foundation and having been so integral to its entire history, will be linked indissolubly to its future by bearing the title "Chairman Emeritus..."

The other contributions to this volume will detail the life and times of the Bell Laboratories during the glory years when it was led by Dr. Baker. I will conclude this brief tribute by saying simply that, for me personally, Bill Baker was an exemplary leader, teacher, and friend.

William G. Bowen
February 9, 2008

EPILOGUE

Michael N. Geselowitz
Staff Director, IEEE History Center

Bell Labs was a very large research and development organization at its peak in the 1960s and 1970s, including station apparatus development, systems engineering, transmission and switching engineering, military systems, and research. The collection of narratives in this book focuses on this time period but only on research, which was led by William O. Baker. Clearly this restricts the breadth of the book, but capturing some history is far better than losing it all and perhaps will stimulate similar documentation of other units within the Labs — and perhaps even other corporate research laboratories. Furthermore, it can be shown that this history has importance beyond the individuals involved, beyond New Jersey, in fact, beyond even corporate or technological history to the history of the United States itself. For almost 90 years, Bell Labs has been at the forefront of technological research and development throughout the world, and contributed greatly to America's leadership in this area.

Of course, like any institution of that duration, it has had its various phases and its ups and downs, especially as its parent company has gone through various organizational forms. Nevertheless, although historians hate to speak in absolutes, an argument can be made that it was the most important research laboratory of the 20th century. Furthermore, although great innovations were born at Bell Labs in every decade from the 1920s through the current day, an argument can be made that the most important period for Bell Labs research was that critical period for the entire world that comprised the "long 1960s" – the height of the Cold War and its technological competition between Superpowers. From the beginning of the space race with the launch of *Sputnik* in 1957 until the signing of the SALT I treaty in 1972, many leaders on both sides of the Cold War thought that the battle would ultimately be won in the laboratories and the factories. Both sides had their corporate labs (after a fashion in the Soviet Union), their university labs, and their government labs. By both its structure and history, however, Bell Labs was unique in the world, and its discoveries and inventions in this period (advances on its earlier invention of the transistor, the laser, UNIX, the charge-coupled device, and so on) were to transform global society and help form both the information age and the digital era. And whether by accident or providence, these years correspond almost exactly with the years when Bill Baker led the Labs (1955 – 1973). Clearly, Baker was building on the earlier structure and triumphs of Bell Labs as he made his mark, but the historian wants to know: Was there something unique about his tenure and his style of management?

A definitive historiography of the Baker period has not been written (let alone of any of the Bell Labs eras, and certainly not of the Labs as a whole). As readers by now know, this slim volume is not it. Rather, it is a contribution to the historical record. The workers at Bell Labs in the Baker years have mostly entered retirement, and we are on the verge of losing their memories (some have already been lost). By collecting the thoughts of an admittedly unique collection of individuals who participated in the Bell Labs story in those years, we are taking the first step toward preserving these voices for future historians.

Therefore, the individual chapters are mini-memoirs, ranging from personal background to research accounts to stories of social life at the Labs. The participants were allowed to maintain their own voice. Some of the anecdotes appear perhaps not as relevant as others for coming to an understanding of what made Bell Labs unique in the Baker years, and future historians may decide what specific aspects — family background, educational background of colleagues, organization of research groups — warrant further study. That is also why individuals from every aspect of the Lab's research operations were included, from chauffeurs and technicians to top scientists. Obviously, we have only included a small sample of the potential participants, but we think that they are representative in many ways of the research organization: they include representatives of every stratum of the Labs operations, from a limo driver to the general patent attorney of the entire Bell Labs. More interestingly, they include a woman, people of different socioeconomic backgrounds, members of families with deep roots in the United States, children of immigrants, and immigrants.

Therefore, even on this thin evidence, I would like to suggest that there are patterns of behavior at Bell Labs in these narratives that give preliminary guidance to future historians studying the origins of its success—and also to future industrial leaders seeking to emulate that success—and I would like to point out five apparently significant factors.

First was the incredible loyalty of these individuals to the institution of Bell Labs. This loyalty was not to the local research group nor to the corporate parent, but to Bell Labs itself. All of these individuals wanted to be there and were proud to be there. The following factors may go a long way to explaining this loyalty, but I believe that future historians should give it careful consideration in its own right.

Second was the attenuation of limitations elsewhere in society. Class or ethnic origin or immigrant status or educational credentials were no barriers to professional or social interaction at the Labs. All that was valued was intelligence and work ethic. Maclennan's narrative does point to some initial struggles by women, but in the end of the day she is sent on a special mission to Antarctica! How many women of her day can make that claim?

Third, in co-editing these contributions and reading them closely as a historian, the idea emerged that this diversity was not just a mechanism to allow a meritocracy to flourish, but actually was itself a condition that led to success. This is a critical thought in the 21st century, when arguments are being made for against the idea that diversity should be a social and political goal perse.

Fourth, there is a management structure where the managers seem to be working for the line researchers rather than the other way around. Again, this may stem from the other factors, but was clearly a critical contributor to the Labs' successes. Business historians—and modern business leaders—will want to give this management style careful review.

Finally, there is Bill Baker himself. His presence runs through all the narratives, either directly or clearly in the background leading the organization and defining its tone. Clearly, he emphasized the four aspects I have mentioned above, and that his personal background, aptitude and leadership increased them as he left an indelible stamp on Bell Labs and, indeed, on global science and technology.

INDEX

Page references in *italics* indicate photographs.

A

ABA (American Bar Association), 291; Patent, Trademark and Copyright (PTC) section, 288
ABM (antiballistic missile) systems, 279
"Acoustic Camera," 89
acoustics research, 80, 125–126; early virtual acoustic images, 68, 92; Philharmonic Hall (New York), 90–93; "Vibrations of Indian Musical Drums Regarded as Composite Membranes" (Ramakrishna and Sondhi), 249;

"Whither Speech Recognition?" (Pierce), 56
ACS (American Chemical Society), 225, 228
Adaptive Predictive Coding (Atal and Schroeder), 88
Advisory Commission on Patent Law Reform by the Secretary of Commerce, 291
aeronautical engineering, 32
AFL/CIO, 290
Agere Systems, 121
AIPPI (International Association for the Protection of Intellectual Property), 302–303
Alcatel, 297
Alcatel-Lucent, 297
Allen, Bob, 300
alpha particles, 113

2007117R00201

Made in the USA
San Bernardino, CA
01 March 2013